Calculus of Variations

I. M. Gelfand
S. V. Fomin

Translated and Edited by
Richard A. Silverman

DOVER PUBLICATIONS, INC.
Mineola, New York

Bibliographical Note

This Dover edition, first published in 2000, is an unabridged reprint of
Calculus of Variations, originally published by Prentice-Hall, Inc.,
Englewood Cliffs, N.J., in 1963. The translator has made minor correc-
tions to the text for this new edition.

Library of Congress Cataloging-in-Publication Data

Gel'fand, I. M. (Izrail' Moiseevich)
  [Variatsionnoe ischislenie. English]
  Calculus of variations / I.M. Gelfand, S.V. Fomin ; translated and
edited by Richard A. Silverman.
    p. cm.
  Originally published: Englewood Cliffs, N.J. : Prentice-Hall, 1963.
  ISBN 0-486-41448-5 (pbk.)
    1. Calculus of variations. I. Fomin, S. V. (Sergei Vasil'evich) II.
Silverman, Richard A. III. Title.

QA315 .G41713 2000
515'.64—dc21

00-057099

Manufactured in the United States of America
Dover Publications, Inc., 31 East 2nd Street, Mineola, N.Y. 11501

AUTHORS' PREFACE

The present course is based on lectures given by I. M. Gelfand in the Mechanics and Mathematics Department of Moscow State University. However, the book goes considerably beyond the material actually presented in the lectures. Our aim is to give a treatment of the elements of the calculus of variations in a form which is both easily understandable and sufficiently modern. Considerable attention is devoted to physical applications of variational methods, e.g., canonical equations, variational principles of mechanics and conservation laws.

The reader who merely wishes to become familiar with the most basic concepts and methods of the calculus of variations need only study the first chapter. The first three chapters, taken together, form a more comprehensive course on the elements of the calculus of variations,—but one which is still quite elementary (involving only necessary conditions for extrema). The first six chapters contain, more or less, the material given in the usual university course in the calculus of variations (with applications to the mechanics of systems with a finite number of degrees of freedom), including the theory of fields (presented in a somewhat novel way) and sufficient conditions for weak and strong extrema. Chapter 7 is devoted to the application of variational methods to the study of systems with infinitely many degrees of freedom. Chapter 8 contains a brief treatment of direct methods in the calculus of variations.

The authors are grateful to M. A. Yevgrafov and A. G. Kostyuchenko, who read the book in manuscript and made many useful comments.

I. M. G.
S. V. F.

TRANSLATOR'S PREFACE

This book is a modern introduction to the calculus of variations and certain of its ramifications, and I trust that its fresh and lively point of view will serve to make it a welcome addition to the English-language literature on the subject. The present edition is rather different from the Russian original. With the authors' consent, I have given free rein to the tendency of any mathematically educated translator to assume the functions of annotator and stylist. In so doing, I have had two special assets: 1) A substantial list of revisions and corrections from Professor S. V. Fomin himself, and 2) A variety of helpful suggestions from Professor J. T. Schwartz of New York University, who read the entire translation in typescript.

The problems appearing at the end of each of the eight chapters and two appendices were made specifically for the English edition, and many of them comment further on the corresponding parts of the text. A variety of Russian sources have played an important role in the synthesis of this material. In particular, I have consulted the textbooks on the calculus of variations by N. I. Akhiezer, by L. E. Elsgolts, and by M. A. Lavrentev and L. A. Lyusternik, as well as Volume 2 of the well-known problem collection by N. M. Gyunter and R. O. Kuzmin, and Chapter 3 of G. E. Shilov's "Mathematical Analysis, A Special Course."

At the end of the book I have added a Bibliography containing suggestions for collateral and supplementary reading. This list is not intended as an exhaustive catalog of the literature, and is in fact confined to books available in English.

R. A. S.

CONTENTS

1

ELEMENTS
OF THE THEORY

I. Functionals. Some Simple Variational Problems

Variable quantities called *functionals* play an important role in many problems arising in analysis, mechanics, geometry, etc. By a *functional*, we mean a correspondence which assigns a definite (real) number to each function (or curve) belonging to some class. Thus, one might say that a functional is a kind of function, where the independent variable is itself a function (or curve). The following are examples of functionals:

1. Consider the set of all rectifiable plane curves.[1] A definite number is associated with each such curve, namely, its length. Thus, the length of a curve is a functional defined on the set of rectifiable curves.

2. Suppose that each rectifiable plane curve is regarded as being made out of some homogeneous material. Then if we associate with each such curve the ordinate of its center of mass, we again obtain a functional.

3. Consider all possible paths joining two given points A and B in the plane. Suppose that a particle can move along any of these paths, and let the particle have a definite velocity $v(x, y)$ at the point (x, y). Then we obtain a functional by associating with each path the time the particle takes to traverse the path.

---

[1] In analysis, the *length* of a curve is defined as the limiting length of a polygonal line inscribed in the curve (i.e., with vertices lying on the curve) as the maximum length of the chords forming the polygonal line goes to zero. If this limit exists and is finite, the curve is said to be *rectifiable*.

4. Let $y(x)$ be an arbitrary continuously differentiable function, defined on the interval $[a, b]$.[2] Then the formula

$$J[y] = \int_a^b y'^2(x)\, dx$$

defines a functional on the set of all such functions $y(x)$.

5. As a more general example, let $F(x, y, z)$ be a continuous function of three variables. Then the expression

$$J[y] = \int_a^b F[x, y(x), y'(x)]\, dx, \qquad (1)$$

where $y(x)$ ranges over the set of all continuously differentiable functions defined on the interval $[a, b]$, defines a functional. By choosing different functions $F(x, y, z)$, we obtain different functionals. For example, if

$$F(x, y, z) = \sqrt{1 + z^2},$$

$J[y]$ is the length of the curve $y = y(x)$, as in the first example, while if

$$F(x, y, z) = z^2,$$

$J[y]$ reduces to the case considered in the fourth example. In what follows, we shall be concerned mainly with functionals of the form (1).

Particular instances of problems involving the concept of a functional were considered more than three hundred years ago, and in fact, the first important results in this area are due to Euler (1707–1783). Nevertheless, up to now, the "calculus of functionals" still does not have methods of a generality comparable to the methods of classical analysis (i.e., the ordinary "calculus of functions"). The most developed branch of the "calculus of functionals" is concerned with finding the maxima and minima of functionals, and is called the "calculus of variations." Actually, it would be more appropriate to call this subject the "calculus of variations in the narrow sense," since the significance of the concept of the variation of a functional is by no means confined to its applications to the problem of determining the extrema of functionals.

We now indicate some typical examples of *variational problems*, by which we mean problems involving the determination of maxima and minima of functionals.

1. *Find the shortest plane curve joining two points A and B, i.e., find the curve $y = y(x)$ for which the functional*

$$\int_a^b \sqrt{1 + y'^2}\, dx$$

achieves its minimum. The curve in question turns out to be the straight line segment joining A and B.

---

[2] By $[a, b]$ is meant the closed interval $a \leqslant x \leqslant b$.

2. Let A and B be two fixed points. Then the time it takes a particle to slide under the influence of gravity along some path joining A and B depends on the choice of the path (curve), and hence is a functional. The curve such that the particle takes the least time to go from A to B is called the *brachistochrone*. The brachistochrone problem was posed by John Bernoulli in 1696, and played an important part in the development of the calculus of variations. The problem was solved by John Bernoulli, James Bernoulli, Newton, and L'Hospital. The brachistochrone turns out to be a cycloid, lying in the vertical plane and passing through A and B (cf. p. 26).

3. The following variational problem, called the *isoperimetric problem,* was solved by Euler: *Among all closed curves of a given length l, find the curve enclosing the greatest area.* The required curve turns out to be a circle.

All of the above problems involve functionals which can be written in the form

$$\int_a^b F(x, y, y') \, dx.$$

Such functionals have a "localization property" consisting of the fact that if we divide the curve $y = y(x)$ into parts and calculate the value of the functional for each part, the sum of the values of the functional for the separate parts equals the value of the functional for the whole curve. It is just these functionals which are usually considered in the calculus of variations. As an example of a "nonlocal functional," consider the expression

$$\frac{\int_a^b x\sqrt{1 + y'^2} \, dx}{\int_a^b \sqrt{1 + y'^2} \, dx},$$

which gives the abscissa of the center of mass of a curve $y = y(x)$, $a \leqslant x \leqslant b$, made out of some homogeneous material.

An important factor in the development of the calculus of variations was the investigation of a number of mechanical and physical problems, e.g., the brachistochrone problem mentioned above. In turn, the methods of the calculus of variations are widely applied in various physical problems. It should be emphasized that the application of the calculus of variations to physics does not consist merely in the solution of individual, albeit very important problems. The so-called "variational principles," to be discussed in Chapters 4 and 7, are essentially a manifestation of very general physical laws, which are valid in diverse branches of physics, ranging from classical mechanics to the theory of elementary particles.

To understand the basic meaning of the problems and methods of the calculus of variations, it is very important to see how they are related to problems of classical analysis, i.e., to the study of functions of n variables. Thus, consider a functional of the form

$$J[y] = \int_a^b F(x, y, y')\, dx, \quad y(a) = A, \quad y(b) = B.$$

Here, each curve is assigned a certain number. To find a related function of the sort considered in classical analysis, we may proceed as follows. Using the points

$$a = x_0, \quad x_1, \ldots, \quad x_n, \quad x_{n+1} = b,$$

we divide the interval $[a, b]$ into $n + 1$ equal parts. Then we replace the curve $y = y(x)$ by the polygonal line with vertices

$$(x_0, A), \quad (x_1, y(x_1)), \ldots, \quad (x_n, y(x_n)), \quad (x_{n+1}, B),$$

and we approximate the functional $J[y]$ by the sum

$$J(y_1, \ldots, y_n) = \sum_{i=1}^{n+1} F\left(x_i, y_i, \frac{y_i - y_{i-1}}{h}\right) h, \tag{2}$$

where

$$y_i = y(x_i), \qquad h = x_i - x_{i-1}.$$

Each polygonal line is uniquely determined by the ordinates $y_1, \ldots, y_n$ of its vertices (recall that $y_0 = A$ and $y_{n+1} = B$ are fixed), and the sum (2) is therefore a function of the n variables $y_1, \ldots, y_n$. Thus, as an approximation, we can regard the variational problem as the problem of finding the extrema of the function $J(y_1, \ldots, y_n)$. In solving variational problems, Euler made extensive use of this *method of finite differences*. By replacing smooth curves by polygonal lines, he reduced the problem of finding extrema of a functional to the problem of finding extrema of a function of n variables, and then he obtained exact solutions by passing to the limit as $n \to \infty$. In this sense, functionals can be regarded as "functions of infinitely many variables" [i.e., the values of the function $y(x)$ at separate points], and the calculus of variations can be regarded as the corresponding analog of differential calculus.

2. Function Spaces

In the study of functions of n variables, it is convenient to use geometric language, by regarding a set of n numbers $(y_1, \ldots, y_n)$ as a point in an n-dimensional space. In just the same way, geometric language is useful when studying functionals. Thus, we shall regard each function $y(x)$ belonging to some class as a point in some space, and spaces whose elements are functions will be called *function spaces*.

In the study of functions of a finite number n of independent variables, it is sufficient to consider a single space, i.e., n-dimensional Euclidean space

$\mathscr{E}_n$.[3] However, in the case of function spaces, there is no such "universal" space. In fact, the nature of the problem under consideration determines the choice of the function space. For example, if we are dealing with a functional of the form

$$\int_a^b F(x, y, y')\, dx,$$

it is natural to regard the functional as defined on the set of all functions with a continuous first derivative, while in the case of a functional of the form

$$\int_a^b F(x, y, y', y'')\, dx,$$

the appropriate function space is the set of all functions with two continuous derivatives. Therefore, in studying functionals of various types, it is reasonable to use various function spaces.

The concept of continuity plays an important role for functionals, just as it does for the ordinary functions considered in classical analysis. In order to formulate this concept for functionals, we must somehow introduce a concept of "closeness" for elements in a function space. This is most conveniently done by introducing the concept of the *norm* of a function, analogous to the concept of the distance between a point in Euclidean space and the origin of coordinates. Although in what follows we shall always be concerned with function spaces, it will be most convenient to introduce the concept of a norm in a more general and abstract form, by introducing the concept of a *normed linear space*.

By a *linear space*, we mean a set $\mathscr{R}$ of elements $x, y, z, \ldots$ of any kind, for which the operations of addition and multiplication by (real) numbers $\alpha, \beta, \ldots$ are defined and obey the following axioms:

1. $x + y = y + x$;
2. $(x + y) + z = x + (y + z)$;
3. There exists an element 0 (the *zero element*) such that $x + 0 = x$ for any $x \in \mathscr{R}$;[4]
4. For each $x \in \mathscr{R}$, there exists an element $-x$ such that $x + (-x) = 0$;
5. $1 \cdot x = x$;
6. $\alpha(\beta x) = (\alpha\beta)x$;
7. $(\alpha + \beta)x = \alpha x + \beta x$;
8. $\alpha(x + y) = \alpha x + \alpha y$.

---

[3] See e.g., G. E. Shilov, *An Introduction to the Theory of Linear Spaces*, translated by R. A. Silverman, Prentice-Hall, Inc., Englewood Cliffs, N. J. (1961), Theorem 14 and Corollary, pp. 48–49.

[4] By $x \in \mathscr{R}$, we mean that the element x belongs to the set $\mathscr{R}$. In these axioms, x, y and z are arbitrary elements of $\mathscr{R}$, while α and β are arbitrary real numbers.

A linear space $\mathscr{R}$ is said to be *normed*, if each element $x \in \mathscr{R}$ is assigned a nonnegative number $\|x\|$, called the *norm* of x, such that

1. $\|x\| = 0$ if and only if $x = 0$;
2. $\|\alpha x\| = |\alpha|\,\|x\|$;
3. $\|x + y\| \leqslant \|x\| + \|y\|$.

In a normed linear space, we can talk about distances between elements, by defining the distance between x and y to be the quantity $\|x - y\|$.

The elements of a normed linear space can be objects of any kind, e.g., numbers, vectors (directed line segments), matrices, functions, etc. The following normed linear spaces are important for our subsequent purposes:

1. The space $\mathscr{C}$, or more precisely $\mathscr{C}(a, b)$, consisting of all continuous functions $y(x)$ defined on a (closed) interval $[a, b]$. By addition of elements of $\mathscr{C}$ and multiplication of elements of $\mathscr{C}$ by numbers, we mean ordinary addition of functions and multiplication of functions by numbers, while the norm is defined as the maximum of the absolute value, i.e.,

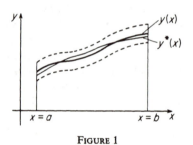

FIGURE 1

$$\|y\|_0 = \max_{a \leqslant x \leqslant b} |y(x)|.$$

   Thus, in the space $\mathscr{C}$, the distance between the function $y^*(x)$ and the function $y(x)$ does not exceed ε if the graph of the function $y^*(x)$ lies inside a strip of width 2ε (in the vertical direction) "bordering" the graph of the function $y(x)$, as shown in Figure 1.

2. The space $\mathscr{D}_1$, or more precisely $\mathscr{D}_1(a, b)$, consisting of all functions $y(x)$ defined on an interval $[a, b]$ which are continuous and have continuous first derivatives. The operations of addition and multiplication by numbers are the same as in $\mathscr{C}$, but the norm is defined by the formula

$$\|y\|_1 = \max_{a \leqslant x \leqslant b} |y(x)| + \max_{a \leqslant x \leqslant b} |y'(x)|.$$

Thus, two functions in $\mathscr{D}_1$ are regarded as close together if both the functions themselves and their first derivatives are close together, since

$$\|y - z\|_1 < \varepsilon$$

implies that

$$|y(x) - z(x)| < \varepsilon, \qquad |y'(x) - z'(x)| < \varepsilon$$

for all $a \leqslant x \leqslant b$.

3. The space $\mathscr{D}_n$, or more precisely $\mathscr{D}_n(a, b)$, consisting of all functions $y(x)$ defined on an interval $[a, b]$ which are continuous and have continuous derivatives up to order n inclusive, where n is a fixed integer. Addition of elements of $\mathscr{D}_n$ and multiplication of elements of $\mathscr{D}_n$ by numbers are defined just as in the preceding cases, but the norm is now defined by the formula

$$\|y\|_n = \sum_{i=0}^{n} \max_{a \leqslant x \leqslant b} |y^{(i)}(x)|,$$

where $y^i(x) = (d/dx)^i y(x)$ and $y^{(0)}(x)$ denotes the function $y(x)$ itself. Thus, two functions in $\mathscr{D}_n$ are regarded as close together if the values of the functions themselves and of all their derivatives up to order n inclusive are close together. It is easily verified that all the axioms of a normed linear space are actually satisfied for each of the spaces $\mathscr{C}$, $\mathscr{D}_1$, and $\mathscr{D}_n$.

Similarly, we can introduce spaces of functions of several variables, e.g., the space of continuous functions of n variables, the space of functions of n variables with continuous first derivatives, etc. After a norm has been introduced in the linear space $\mathscr{R}$ (which may be a function space), it is natural to talk about continuity of functionals defined on $\mathscr{R}$:

DEFINITION. *The functional $J[y]$ is said to be continuous at the point $\hat{y} \in \mathscr{R}$ if for any $\varepsilon > 0$, there is a $\delta > 0$ such that*

$$|J[y] - J[\hat{y}]| < \varepsilon, \tag{3}$$

provided that $\|y - \hat{y}\| < \delta$.

Remark 1. The inequality (3) is equivalent to the two inequalities

$$J[y] - J[\hat{y}] > -\varepsilon \tag{4}$$

and

$$J[y] - J[\hat{y}] < \varepsilon. \tag{5}$$

If in the definition of continuity, we replace (3) by (4), $J[y]$ is said to be *lower semicontinuous* at $\hat{y}$, while if we replace (3) by (5), $J[y]$ is said to be *upper semicontinuous* at $\hat{y}$. These concepts will be needed in Chapter 8.

Remark 2. At first, it might appear that the space $\mathscr{C}$, which is the largest of those enumerated, would be adequate for the study of variational problems. However, this is not the case. In fact, as already mentioned, one of the basic types of functionals considered in the calculus of variations has the form

$$J[y] = \int_a^b F(x, y, y') \, dx.$$

It is easy to see that such a functional (e.g., arc length) will be continuous if we interpret closeness of functions as closeness in the space $\mathscr{D}_1$. However, in general, the functional will not be continuous if we use the norm introduced in the space $\mathscr{C}$,[5] even though it is continuous in the norm of the space $\mathscr{D}_1$. Since we want to be able to use ordinary analytic methods, e.g., passage to the limit, then, given a functional, it is reasonable to choose a function space such that the functional is continuous.

Remark 3. So far, we have talked about linear spaces and functionals defined on them. However, in many variational problems, we have to deal with functionals defined on sets of functions which do not form linear spaces. In fact, the set of functions (or curves) satisfying the constraints of a given variational problem, called the *admissible functions* (or *admissible curves*), is in general not a linear space. For example, the admissible curves for the "simplest" variational problem (see Sec. 4) are the smooth plane curves passing through two fixed points, and the sum of two such curves does not pass through the two points. Nevertheless, the concept of a normed linear space and the related concepts of the distance between functions, continuity of functionals, etc., play an important role in the calculus of variations. A similar situation is encountered in elementary analysis, where, in dealing with functions of n variables, it is convenient to use the concept of an n-dimensional Euclidean space $\mathscr{E}_n$, even though the domain of definition of a function may not be a linear subspace of $\mathscr{E}_n$.

3. The Variation of a Functional. A Necessary Condition for an Extremum

3.1. In this section, we introduce the concept of the *variation* (or *differential*) of a functional, analogous to the concept of the differential of a function of n variables. The concept will then be used to find extrema of functionals. First, we give some preliminary facts and definitions.

DEFINITION. *Given a normed linear space $\mathscr{R}$, let each element $h \in \mathscr{R}$ be assigned a number $\varphi[h]$, i.e., let $\varphi[h]$ be a functional defined on $\mathscr{R}$. Then $\varphi[h]$ is said to be a (continuous) linear functional if*

1. $\varphi[\alpha h] = \alpha \varphi[h]$ *for any $h \in \mathscr{R}$ and any real number α;*
2. $\varphi[h_1 + h_2] = \varphi[h_1] + \varphi[h_2]$ *for any $h_1, h_2 \in \mathscr{R}$;*
3. $\varphi[h]$ *is continuous (for all $h \in \mathscr{R}$).*

Example 1. If we associate with each function $h(x) \in \mathscr{C}(a, b)$ its value at a fixed point x_0 in $[a, b]$, i.e., if we define the functional $\varphi[h]$ by the formula

$$\varphi[h] = h(x_0),$$

then $\varphi[h]$ is a linear functional on $\mathscr{C}(a, b)$.

---

[5] Arc length is a typical example of such a functional. For every curve, we can find another curve arbitrarily close to the first in the sense of the norm of the space $\mathscr{C}$, whose length differs from that of the first curve by a factor of 10, say.

Example 2. The integral

$$\varphi[h] = \int_a^b h(x)\,dx$$

defines a linear functional on $\mathscr{C}(a, b)$.

Example 3. The integral

$$\varphi[h] = \int_a^b \alpha(x)h(x)\,dx,$$

where $\alpha(x)$ is a fixed function in $\mathscr{C}(a, b)$, defines a linear functional on $\mathscr{C}(a, b)$.

Example 4. More generally, the integral

$$\varphi[h] = \int_a^b [\alpha_0(x)h(x) + \alpha_1(x)h'(x) + \cdots + \alpha_n(x)h^{(n)}(x)]\,dx, \tag{6}$$

where the $\alpha_i(x)$ are fixed functions in $\mathscr{C}(a, b)$, defines a linear functional on $\mathscr{D}_n(a, b)$.

Suppose the linear functional (6) vanishes for all $h(x)$ belonging to some class. Then what can be said about the functions $\alpha_i(x)$? Some typical results in this direction are given by the following lemmas:

LEMMA 1. *If $\alpha(x)$ is continuous in $[a, b]$, and if*

$$\int_a^b \alpha(x)h(x)\,dx = 0$$

for every function $h(x) \in \mathscr{C}(a, b)$ such that $h(a) = h(b) = 0$, then $\alpha(x) = 0$ for all x in $[a, b]$.

Proof. Suppose the function $\alpha(x)$ is nonzero, say positive, at some point in $[a, b]$. Then $\alpha(x)$ is also positive in some interval $[x_1, x_2]$ contained in $[a, b]$. If we set

$$h(x) = (x - x_1)(x_2 - x)$$

for x in $[x_1, x_2]$ and $h(x) = 0$ otherwise, then $h(x)$ obviously satisfies the conditions of the lemma. However,

$$\int_a^b \alpha(x)h(x)\,dx = \int_{x_1}^{x_2} \alpha(x)(x - x_1)(x_2 - x)\,dx > 0,$$

since the integrand is positive (except at x_1 and x_2). This contradiction proves the lemma.

Remark. The lemma still holds if we replace $\mathscr{C}(a, b)$ by $\mathscr{D}_n(a, b)$. To see this, we use the same proof with

$$h(x) = [(x - x_1)(x_2 - x)]^{n+1}$$

for x in $[x_1, x_2]$ and $h(x) = 0$ otherwise.

LEMMA 2. *If $\alpha(x)$ is continuous in $[a, b]$, and if*

$$\int_a^b \alpha(x)h'(x)\, dx = 0$$

for every function $h(x) \in \mathscr{D}_1(a, b)$ such that $h(a) = h(b) = 0$, then $\alpha(x) = c$ for all x in $[a, b]$, where c is a constant.

Proof. Let c be the constant defined by the condition

$$\int_a^b [\alpha(x) - c]\, dx = 0,$$

and let

$$h(x) = \int_a^x [\alpha(\xi) - c]\, d\xi,$$

so that $h(x)$ automatically belongs to $\mathscr{D}_1(a, b)$ and satisfies the conditions $h(a) = h(b) = 0$. Then on the one hand,

$$\int_a^b [\alpha(x) - c]h'(x)\, dx = \int_a^b \alpha(x)h'(x)\, dx - c[h(b) - h(a)] = 0,$$

while on the other hand,

$$\int_a^b [\alpha(x) - c]h'(x)\, dx = \int_a^b [\alpha(x) - c]^2\, dx.$$

It follows that $\alpha(x) - c = 0$, i.e., $\alpha(x) = c$, for all x in $[a, b]$.

The next lemma will be needed in Chapter 8:

LEMMA 3. *If $\alpha(x)$ is continuous in $[a, b]$, and if*

$$\int_a^b \alpha(x)h''(x)\, dx = 0$$

for every function $h(x) \in \mathscr{D}_2(a, b)$ such that $h(a) = h(b) = 0$ and $h'(a) = h'(b) = 0$, then $\alpha(x) = c_0 + c_1 x$ for all x in $[a, b]$, where c_0 and c_1 are constants.

Proof. Let c_0 and c_1 be defined by the conditions

$$\int_a^b [\alpha(x) - c_0 - c_1 x]\, dx = 0,$$
$$\int_a^b dx \int_a^x [\alpha(\xi) - c_0 - c_1 \xi]\, d\xi = 0, \tag{7}$$

and let

$$h(x) = \int_a^x d\xi \int_a^\xi [\alpha(t) - c_0 - c_1 t]\, dt,$$

so that $h(x)$ automatically belongs to $\mathscr{D}_2(a, b)$ and satisfies the conditions $h(a) = h(b) = 0$, $h'(a) = h'(b) = 0$. Then on the one hand,

$$\int_a^b [\alpha(x) - c_0 - c_1 x]h''(x)\, dx$$
$$= \int_a^b \alpha(x)h''(x)\, dx - c_0[h'(b) - h'(a)] - c_1 \int_a^b xh''(x)\, dx$$
$$= -c_1[bh'(b) - ah'(a)] - c_1[h(b) - h(a)] = 0,$$

while on the other hand,

$$\int_a^b [\alpha(x) - c_0 - c_1 x] h''(x) \, dx = \int_a^b [\alpha(x) - c_0 - c_1 x]^2 \, dx = 0.$$

It follows that $\alpha(x) - c_0 - c_1 x = 0$, i.e., $\alpha(x) = c_0 + c_1 x$, for all x in $[a, b]$.

LEMMA 4. *If $\alpha(x)$ and $\beta(x)$ are continuous in $[a, b]$, and if*

$$\int_a^b [\alpha(x)h(x) + \beta(x)h'(x)] \, dx = 0 \qquad (8)$$

for every function $h(x) \in \mathscr{D}_1(a, b)$ such that $h(a) = h(b) = 0$, then $\beta(x)$ is differentiable, and $\beta'(x) = \alpha(x)$ for all x in $[a, b]$.

Proof. Setting

$$A(x) = \int_a^x \alpha(\xi) \, d\xi,$$

and integrating by parts, we find that

$$\int_a^b \alpha(x)h(x) \, dx = - \int_a^b A(x)h'(x) \, dx,$$

i.e., (8) can be rewritten as

$$\int_a^b [-A(x) + \beta(x)]h'(x) \, dx = 0.$$

But, according to Lemma 2, this implies that

$$\beta(x) - A(x) = \text{const},$$

and hence by the definition of $A(x)$,

$$\beta'(x) = \alpha(x),$$

for all x in $[a, b]$, as asserted. We emphasize that the differentiability of the function $\beta(x)$ was not assumed in advance.

3.2. We now introduce the concept of the *variation* (or *differential*) of a functional. Let $J[y]$ be a functional defined on some normed linear space, and let

$$\Delta J[h] = J[y + h] - J[y]$$

be its *increment*, corresponding to the increment $h = h(x)$ of the "independent variable" $y = y(x)$. If y is fixed, $\Delta J[h]$ is a functional of h, in general a nonlinear functional. Suppose that

$$\Delta J[h] = \varphi[h] + \varepsilon \|h\|,$$

where $\varphi[h]$ is a linear functional and $\varepsilon \to 0$ as $\|h\| \to 0$. Then the functional $J[y]$ is said to be *differentiable*, and the *principal linear part* of the increment

$\Delta J[h]$, i.e., the linear functional $\varphi[h]$ which differs from $\Delta J[h]$ by an infinitesimal of order higher than 1 relative to $\|h\|$, is called the *variation* (or *differential*) of $J[y]$ and is denoted by $\delta J[h]$.[6]

THEOREM 1.   *The differential of a differentiable functional is unique.*

Proof.   First, we note that if $\varphi[h]$ is a linear functional and if

$$\frac{\varphi[h]}{\|h\|} \to 0$$

as $\|h\| \to 0$, then $\varphi[h] \equiv 0$, i.e., $\varphi[h] = 0$ for all h. In fact, suppose $\varphi[h_0] \neq 0$ for some $h_0 \neq 0$. Then, setting

$$h_n = \frac{h_0}{n}, \qquad \lambda = \frac{\varphi[h_0]}{\|h_0\|},$$

we see that $\|h_n\| \to 0$ as $n \to \infty$, but

$$\lim_{n \to \infty} \frac{\varphi[h_n]}{\|h_n\|} = \lim_{n \to \infty} \frac{n\varphi[h_0]}{n\|h_0\|} = \lambda \neq 0,$$

contrary to hypothesis.

Now, suppose the differential of the functional $J[y]$ is not uniquely defined, so that

$$\Delta J[h] = \varphi_1[h] + \varepsilon_1\|h\|,$$
$$\Delta J[h] = \varphi_2[h] + \varepsilon_2\|h\|,$$

where $\varphi_1[h]$ and $\varphi_2[h]$ are linear functionals, and $\varepsilon_1, \varepsilon_2 \to 0$ as $\|h\| \to 0$. This implies

$$\varphi_1[h] - \varphi_2[h] = \varepsilon_2\|h\| - \varepsilon_1\|h\|,$$

and hence $\varphi_1[h] - \varphi_2[h]$ is an infinitesimal of order higher than 1 relative to $\|h\|$. But since $\varphi_1[h] - \varphi_2[h]$ is a linear functional, it follows from the first part of the proof that $\varphi_1[h] - \varphi_2[h]$ vanishes identically, as asserted.

Next, we use the concept of the variation (or) differential of a functional to establish a necessary condition for a functional to have an extremum. We begin by recalling the corresponding concepts from analysis. Let $F(x_1, \ldots, x_n)$ be a differentiable function of n variables. Then $F(x_1, \ldots, x_n)$ is said to have a *(relative) extremum* at the point $(\hat{x}_1, \ldots, \hat{x}_n)$ if

$$\Delta F = F(x_1, \ldots, x_n) - F(\hat{x}_1, \ldots, \hat{x}_n)$$

has the same sign for all points $(x_1, \ldots, x_n)$ belonging to some neighborhood of $(\hat{x}_1, \ldots, \hat{x}_n)$, where the extremum $F(\hat{x}_1, \ldots, \hat{x}_n)$ is a *minimum* if $\Delta F \geqslant 0$ and a *maximum* if $\Delta F \leqslant 0$.

Analogously, we say that the functional $J[y]$ has a *(relative) extremum* for $y = \hat{y}$ if $J[y] - J[\hat{y}]$ does not change its sign in some neighborhood of

---

[6] Strictly speaking, of course, the increment and the variation of $J[y]$, are functionals of two arguments y and h, and to emphasize this fact, we might write $\Delta J[y; h] = \delta J[y; h] + \varepsilon\|h\|$.

the curve $y = \hat{y}(x)$. Subsequently, we shall be concerned with functionals defined on some set of continuously differentiable functions, and the functions themselves can be regarded either as elements of the space $\mathscr{C}$ or elements of the space $\mathscr{D}_1$. Corresponding to these two possibilities, we can define two kinds of extrema: We shall say that the functional $J[y]$ has a *weak extremum* for $y = \hat{y}$ if there exists an $\varepsilon > 0$ such that $J[y] - J[\hat{y}]$ has the same sign for all y in the domain of definition of the functional which satisfy the condition $\|y - \hat{y}\|_1 < \varepsilon$, where $\| \ \|_1$ denotes the norm in the space $\mathscr{D}_1$. On the other hand, we shall say that the functional $J[y]$ has a *strong extremum* for $y = \hat{y}$ if there exists an $\varepsilon > 0$ such that $J[y] - J[\hat{y}]$ has the same sign for all y in the domain of definition of the functional which satisfy the condition $\|y - \hat{y}\|_0 < \varepsilon$, where $\| \ \|_0$ denotes the norm in the space $\mathscr{C}$. It is clear that every strong extremum is simultaneously a weak extremum, since if $\|y - \hat{y}\|_1 < \varepsilon$, then $\|y - \hat{y}\|_0 < \varepsilon$, *a fortiori*, and hence, if $J[\hat{y}]$ is an extremum with respect to all y such that $\|y - \hat{y}\|_0 < \varepsilon$, then $J[\hat{y}]$ is certainly an extremum with respect to all y such that $\|y - \hat{y}\|_1 < \varepsilon$. However, the converse is not true in general, i.e., a weak extremum may not be a strong extremum. As a rule, finding a weak extremum is simpler than finding a strong extremum. The reason for this is that the functionals usually considered in the calculus of variations are continuous in the norm of the space $\mathscr{D}_1$ (as noted at the end of the previous section), and this continuity can be exploited in the theory of weak extrema. In general, however, our functionals will not be continuous in the norm of the space $\mathscr{C}$.

THEOREM 2. *A necessary condition for the differentiable functional $J[y]$ to have an extremum for $y = \hat{y}$ is that its variation vanish for $y = \hat{y}$, i.e., that*

$$\delta J[h] = 0$$

for $y = \hat{y}$ and all admissible h.

Proof. To be explicit, suppose $J[y]$ has a minimum for $y = \hat{y}$. According to the definition of the variation $\delta J[h]$, we have

$$\Delta J[h] = \delta J[h] + \varepsilon \|h\|, \tag{9}$$

where $\varepsilon \to 0$ as $\|h\| \to 0$. Thus, for sufficiently small $\|h\|$, the sign of $\Delta J[h]$ will be the same as the sign of $\delta J[h]$. Now, suppose that $\delta J[h_0] \neq 0$ for some admissible h_0. Then for any $\alpha > 0$, no matter how small, we have

$$\delta J[-\alpha h_0] = -\delta J[\alpha h_0].$$

Hence, (9) can be made to have either sign for arbitrarily small $\|h\|$. But this is impossible, since by hypothesis $J[y]$ has a minimum for $y = \hat{y}$, i.e.,

$$\Delta J[h] = J[\hat{y} + h] - J[\hat{y}] \geqslant 0$$

for all sufficiently small $\|h\|$. This contradiction proves the theorem.

Remark. In elementary analysis, it is proved that for a function to have a minimum, it is necessary not only that its first differential vanish ($df = 0$), but also that its second differential be nonnegative. Consideration of the analogous problem for functionals will be postponed until Chapter 5.

4. The Simplest Variational Problem.   Euler's Equation

4.1. We begin our study of concrete variational problems by considering what might be called the "simplest" variational problem, which can be formulated as follows: *Let $F(x, y, z)$ be a function with continuous first and second (partial) derivatives with respect to all its arguments. Then, among all functions $y(x)$ which are continuously differentiable for $a \leqslant x \leqslant b$ and satisfy the boundary conditions*

$$y(a) = A, \qquad y(b) = B, \tag{10}$$

find the function for which the functional

$$J[y] = \int_a^b F(x, y, y') \, dx \tag{11}$$

has a weak extremum. In other words, the simplest variational problem consists of finding a weak extremum of a functional of the form (11), where the class of admissible curves (see p. 8) consists of all smooth curves joining two points. The first two examples on pp. 2, 3, involving the brachistochrone and the shortest distance between two points, are variational problems of just this type. To apply the necessary condition for an extremum (found in Sec. 3.2) to the problem just formulated, we have to be able to calculate the variation of a functional of the type (11). We now derive the appropriate formula for this variation.

Suppose we give $y(x)$ an increment $h(x)$, where, in order for the function

$$y(x) + h(x)$$

to continue to satisfy the boundary conditions, we must have

$$h(a) = h(b) = 0.$$

Then, since the corresponding increment of the functional (11) equals

$$\Delta J = J[y + h] - J[y] = \int_a^b F(x, y + h, y' + h') \, dx - \int_a^b F(x, y, y') \, dx$$
$$= \int_a^b [F(x, y + h, y' + h') - F(x, y, y')] \, dx,$$

it follows by using Taylor's theorem that

$$\Delta J = \int_a^b [F_y(x, y, y')h + F_{y'}(x, y, y')h'] \, dx + \cdots, \tag{12}$$

where the subscripts denote partial derivatives with respect to the corresponding arguments, and the dots denote terms of order higher than 1 relative to h and h'. The integral in the right-hand side of (12) represents the principal linear part of the increment ΔJ, and hence the variation of $J[y]$ is

$$\delta J = \int_a^b [F_y(x, y, y')h + F_{y'}(x, y, y')h'] \, dx.$$

According to Theorem 2 of Sec. 3.2, a necessary condition for $J[y]$ to have an extremum for $y = y(x)$ is that

$$\delta J = \int_a^b (F_y h + F_{y'} h') \, dx = 0 \tag{13}$$

for all admissible h. But according to Lemma 4 of Sec. 3.1, (13) implies that

$$F_y - \frac{d}{dx} F_{y'} = 0, \tag{14}$$

a result known as *Euler's equation*.[7]   Thus, we have proved

THEOREM 1. *Let $J[y]$ be a functional of the form*

$$\int_a^b F(x, y, y') \, dx,$$

defined on the set of functions $y(x)$ which have continuous first derivatives in $[a, b]$ and satisfy the boundary conditions $y(a) = A$, $y(b) = B$. Then a necessary condition for $J[y]$ to have an extremum for a given function $y(x)$ is that $y(x)$ satisfy Euler's equation[8]

$$F_y - \frac{d}{dx} F_{y'} = 0.$$

The integral curves of Euler's equation are called *extremals*. Since Euler's equation is a second-order differential equation, its solution will in general depend on two arbitrary constants, which are determined from the boundary conditions $y(a) = A$, $y(b) = B$. The problem usually considered in the theory of differential equations is that of finding a solution which is defined in the neighborhood of some point and satisfies given initial conditions (*Cauchy's problem*). However, in solving Euler's equation, we are looking for a solution which is defined over all of some fixed region and satisfies given boundary conditions. Therefore, the question of whether or not a certain variational problem has a solution does not just reduce to the

---

[7] We emphasize that the existence of the derivative $(d/dx)F_{y'}$ is not assumed in advance, but follows from the very same lemma.

[8] This condition is necessary for a weak extremum. Since every strong extremum is simultaneously a weak extremum, any necessary condition for a weak extremum is also a necessary condition for a strong extremum.

usual existence theorems for differential equations.   In this regard, we now state a theorem due to Bernstein,[9] concerning the existence and uniqueness of solutions "in the large" of an equation of the form

$$y'' = F(x, y, y'). \tag{15}$$

THEOREM 2 (*Bernstein*).   *If the functions F, F_y and $F_{y'}$ are continuous at every finite point (x, y) for any finite y', and if a constant $k > 0$ and functions*

$$\alpha = \alpha(x, y) \geqslant 0, \qquad \beta = \beta(x, y) \geqslant 0$$

(which are bounded in every finite region of the plane) can be found such that

$$F_y(x, y, y') > k, \qquad |F(x, y, y')| \leqslant \alpha y'^2 + \beta,$$

then one and only one integral curve of equation (15) *passes through any two points* (a, A) *and* (b, B) *with different abscissas* $(a \neq b)$.

Equation (13) gives a necessary condition for an extremum, but in general, one which is not sufficient.   The question of sufficient conditions for an extremum will be considered in Chapter 5.   In many cases, however, Euler's equation by itself is enough to give a complete solution of the problem.   In fact, the existence of an extremum is often clear from the physical or geometric meaning of the problem, e.g., in the brachistochrone problem, the problem concerning the shortest distance between two points, etc.   If in such a case there exists only one extremal satisfying the boundary conditions of the problem, this extremal must perforce be the curve for which the extremum is achieved.

For a functional of the form

$$\int_a^b F(x, y, y')\, dx$$

Euler's equation is in general a second-order differential equation, but it may turn out that the curve for which the functional has its extremum is not twice differentiable.   For example, consider the functional

$$J[y] = \int_{-1}^1 y^2 (2x - y')^2\, dx,$$

where

$$y(-1) = 0, \qquad y(1) = 1.$$

The minimum of $J[y]$ equals zero and is achieved for the function

$$y = y(x) = \begin{cases} 0 & \text{for} \quad -1 \leqslant x \leqslant 0, \\ x^2 & \text{for} \quad\ \ 0 < x \leqslant 1, \end{cases}$$

---

[9] S. N. Bernstein, *Sur les équations du calcul des variations*, Ann. Sci. École Norm. Sup., **29**, 431–485 (1912).

which has no second derivative for $x = 0$. Nevertheless, $y(x)$ satisfies the appropriate Euler equation. In fact, since in this case

$$F(x, y, y') = y^2(2x - y')^2,$$

it follows that all the functions

$$F_y = 2y(2x - y')^2, \quad F_{y'} = -2y^2(2x - y'), \quad \frac{d}{dx}F_{y'}$$

vanish identically for $-1 \leqslant x \leqslant 1$. Thus, despite the fact that Euler's equation is of the second order and $y''(x)$ does not exist everywhere in $[-1, 1]$, substitution of $y(x)$ into Euler's equation converts it into an identity.

We now give conditions guaranteeing that a solution of Euler's equation has a second derivative:

THEOREM 3. *Suppose $y = y(x)$ has a continuous first derivative and satisfies Euler's equation*

$$F_y - \frac{d}{dx}F_{y'} = 0.$$

Then, if the function $F(x, y, y')$ has continuous first and second derivatives with respect to all its arguments, $y(x)$ has a continuous second derivative at all points (x, y) where

$$F_{y'y'}[x, y(x), y'(x)] \neq 0.$$

Proof. Consider the difference

$$\begin{aligned} \Delta F_{y'} &= F_{y'}(x + \Delta x, y + \Delta y, y' + \Delta y') - F_{y'}(x, y, y') \\ &= \Delta x \bar{F}_{y'x} + \Delta y \bar{F}_{y'y} + \Delta y' \bar{F}_{y'y'}, \end{aligned}$$

where the overbar indicates that the corresponding derivatives are evaluated along certain intermediate curves. We divide this difference by Δx, and consider the limit of the resulting expression

$$\bar{F}_{y'x} + \frac{\Delta y}{\Delta x}\bar{F}_{y'y} + \frac{\Delta y'}{\Delta x}\bar{F}_{y'y'}$$

as $\Delta x \to 0$. (This limit exists, since $F_{y'}$ has a derivative with respect to x, which, according to Euler's equation, equals F_y.) Since, by hypothesis, the second derivatives of $F(x, y, z)$ are continuous, then, as $\Delta x \to 0$, $\bar{F}_{y'x}$ converges to $F_{y'x}$, i.e., to the value of $\partial^2 F/\partial y' \, \partial x$ at the point x. It follows from the existence of y' and the continuity of the second derivative $F_{y'y}$ that the second term $(\Delta y/\Delta x)\bar{F}_{y'y}$ also has a limit as $\Delta x \to 0$. But then the third term also has a limit (since the limit of the sum of the three terms exists), i.e., the limit

$$\lim_{\Delta x \to 0} \frac{\Delta y'}{\Delta x}\bar{F}_{y'y'}$$

exists.   As $\Delta x \to 0$, $\bar{F}_{y'y'}$ converges to $F_{y'y'} \neq 0$, and hence

$$\lim_{\Delta x \to 0} \frac{\Delta y'}{\Delta x} = y''(x)$$

exists.   Finally, from the equation

$$\frac{d}{dx} F_{y'} - F_y = 0,$$

we can find an expression for y'', from which it is clear that y'' is continuous wherever $F_{y'y'} \neq 0$.   This proves the theorem.

Remark.   Here it is assumed that the extremals are *smooth*.[10]   In Sec. 15 we shall consider the case where the solution of a variational problem may only be *piecewise smooth*, i.e., may have "corners" at certain points.

4.2. Euler's equation (14) plays a fundamental role in the calculus of variations, and is in general a second-order differential equation.   We now indicate some special cases where Euler's equation can be reduced to a first-order differential equation, or where its solution can be obtained entirely in terms of quadratures (i.e., by evaluating integrals).

Case 1. Suppose the integrand does not depend on y, i.e., let the functional under consideration have the form

$$\int_a^b F(x, y') \, dx,$$

where F does not contain y explicitly.   In this case, Euler's equation becomes

$$\frac{d}{dx} F_{y'} = 0,$$

which obviously has the first integral

$$F_{y'} = C, \tag{16}$$

where C is a constant.   This is a first-order differential equation which does not contain y.   Solving (16) for y', we obtain an equation of the form

$$y' = f(x, C),$$

from which y can be found by a quadrature.

Case 2. If the integrand does not depend on x, i.e., if

$$J[y] = \int_a^b F(y, y') \, dx,$$

then

$$F_y - \frac{d}{dx} F_{y'} = F_y - F_{y'y} y' - F_{y'y'} y''. \tag{17}$$

---

[10] We say that the function $y(x)$ is *smooth* in an interval $[a, b]$ if it is continuous in $[a, b]$, and has a continuous derivative in $[a, b]$.   We say that $y(x)$ is *piecewise smooth* in $[a, b]$ if it is continuous everywhere in $[a, b]$, and has a continuous derivative in $[a, b]$ except possibly at a finite number of points.

Multiplying (17) by y', we obtain

$$F_y y' - F_{y'y} y'^2 - F_{y'y'} y' y'' = \frac{d}{dx}(F - y'F_{y'}).$$

Thus, in this case, Euler's equation has the first integral

$$F - y'F_{y'} = C,$$

where C is a constant.

Case 3. *If F does not depend on y'*, Euler's equation takes the form

$$F_y(x, y) = 0,$$

and hence is not a differential equation, but a "finite" equation, whose solution consists of one or more curves $y = y(x)$.

Case 4. In a variety of problems, one encounters functionals of the form

$$\int_a^b f(x, y)\sqrt{1 + y'^2}\, dx,$$

representing the integral of a function $f(x, y)$ with respect to the *arc length* s $(ds = \sqrt{1 + y'^2}\, dx)$. In this case, Euler's equation can be transformed into

$$\frac{\partial F}{\partial y} - \frac{d}{dx}\left(\frac{\partial F}{\partial y'}\right) = f_y(x, y)\sqrt{1 + y'^2} - \frac{d}{dx}\left[f(x, y)\frac{y'}{\sqrt{1 + y'^2}}\right]$$

$$= f_y\sqrt{1 + y'^2} - f_x\frac{y'}{\sqrt{1 + y'^2}} - f_y\frac{y'^2}{\sqrt{1 + y'^2}} - f\frac{y''}{(1 + y'^2)^{3/2}}$$

$$= \frac{1}{\sqrt{1 + y'^2}}\left[f_y - f_x y' - f\frac{y''}{1 + y'^2}\right] = 0,$$

i.e.,

$$f_y - f_x y' - f\frac{y''}{1 + y'^2} = 0.$$

Example 1. Suppose that

$$J[y] = \int_1^2 \frac{\sqrt{1 + y'^2}}{x}\, dx, \quad y(1) = 0, \quad y(2) = 1.$$

The integrand does not contain y, and hence Euler's equation has the form $F_{y'} = C$ (cf. Case 1). Thus,

$$\frac{y'}{x\sqrt{1 + y'^2}} = C,$$

so that

$$y'^2(1 - C^2 x^2) = C^2 x^2$$

or

$$y' = \frac{Cx}{\sqrt{1 - C^2 x^2}},$$

from which it follows that

$$y = \int \frac{Cx\,dx}{\sqrt{1 - C^2x^2}} = \frac{1}{C}\sqrt{1 - C^2x^2} + C_1$$

or

$$(y - C_1)^2 + x^2 = \frac{1}{C^2}.$$

Thus, the solution is a circle with its center on the y-axis. From the conditions $y(1) = 0$, $y(2) = 1$, we find that

$$C = \frac{1}{\sqrt{5}}, \qquad C_1 = 2,$$

so that the final solution is

$$(y - 2)^2 + x^2 = 5.$$

Example 2. *Among all the curves joining two given points (x_0, y_0) and (x_1, y_1), find the one which generates the surface of minimum area when rotated about the x-axis.* As we know, the area of the surface of revolution generated by rotating the curve $y = y(x)$ about the x-axis is

$$2\pi \int_{x_0}^{x_1} y\sqrt{1 + y'^2}\,dx.$$

Since the integrand does not depend explicitly on x, Euler's equation has the first integral

$$F - y'F_{y'} = C$$

(cf. Case 2), i.e.,

$$y\sqrt{1 + y'^2} - y\frac{y'^2}{\sqrt{1 + y'^2}} = C$$

or

$$y = C\sqrt{1 + y'^2},$$

so that

$$y' = \sqrt{\frac{y^2 - C^2}{C^2}}.$$

Separating variables, we obtain

$$dx = \frac{C\,dy}{\sqrt{y^2 - C^2}},$$

i.e.,

$$x + C_1 = C \ln \frac{y + \sqrt{y^2 - C^2}}{C},$$

so that

$$y = C \cosh \frac{x + C_1}{C}. \tag{18}$$

Thus, the required curve is a *catenary* passing through the two given points.   The surface generated by rotation of the catenary is called a *catenoid*. The values of the arbitrary constants C and C_1 are determined by the conditions

$$y(x_0) = y_0, \qquad y(x_1) = y_1.$$

It can be shown that the following three cases are possible, depending on the positions of the points (x_0, y_0) and (x_1, y_1):

1. If a single curve of the form (18) can be drawn through the points (x_0, y_0) and (x_1, y_1), this curve is the solution of the problem [see Figure 2(a)].

2. If two extremals can be drawn through the points (x_0, y_0) and (x_1, y_1), one of the curves actually corresponds to the surface of revolution of minimum area, and the other does not.

3. If there is no curve of the form (18) passing through the points (x_0, y_0) and (x_1, y_1), there is no surface in the class of smooth surfaces of revolution which achieves the minimum area.   In fact, if the location of the

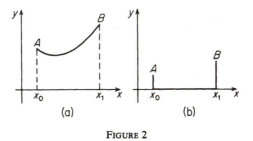

FIGURE 2

two points is such that the distance between them is sufficiently large compared to their distances from the x-axis, then the area of the surface consisting of two circles of radius y_0 and y_1, plus the segment of the x-axis joining them [see Figure 2(b)] will be less than the area of any surface of revolution generated by a smooth curve passing through the points.   Thus, in this case the surface of revolution generated by the polygonal line Ax_0x_1B has the minimum area, and there is no surface of minimum area in the class of surfaces generated by rotation about the x-axis of smooth curves passing through the given points.   (This case, corresponding to a "broken extremal," will be discussed further in Sec. 15.)

Example 3. For the functional

$$J[y] = \int_a^b (x - y)^2 \, dx, \tag{19}$$

Euler's equation reduces to a finite equation (see Case 3), whose solution is the straight line $y = x$. In fact, the integral (19) vanishes along this line.

5. The Case of Several Variables

So far, we have considered functionals depending on functions of one variable, i.e., on curves. In many problems, however, one encounters functionals depending on functions of several independent variables, i.e., on surfaces. Such multidimensional problems will be considered in detail in Chapter 7. For the time being, we merely give an idea of how the formulation and solution of the simplest variational problem discussed above carries over to the case of functionals depending on surfaces.

To keep the notation simple, we confine ourselves to the case of two independent variables, but all our considerations remain the same when there are n independent variables. Thus, let $F(x, y, z, p, q)$ be a function with continuous first and second (partial) derivatives with respect to all its arguments, and consider a functional of the form

$$J[z] = \int\int_R F(x, y, z, z_x, z_y)\, dx\, dy, \tag{20}$$

where R is some closed region and z_x, z_y are the partial derivatives of $z = z(x, y)$. Suppose we are looking for a function $z(x, y)$ such that

1. $z(x, y)$ and its first and second derivatives are continuous in R;
2. $z(x, y)$ takes given values on the boundary Γ of R;
3. The functional (20) has an extremum for $z = z(x, y)$.

Since the proof of Theorem 2 of Sec. 3.2 does not depend on the form of the functional J, then, just as in the case of one variable, a necessary condition for the functional (20) to have an extremum is that its variation (i.e., the principal linear part of its increment) vanish. However, to find Euler's equation for the functional (20), we need the following lemma, which is analogous to Lemma 1 of Sec. 3.1 (see also the remark on p. 9):

LEMMA. *If* $\alpha(x, y)$ *is a fixed function which is continuous in a closed region R, and if the integral*

$$\int\int_R \alpha(x, y)h(x, y)\, dx\, dy \tag{21}$$

vanishes for every function $h(x, y)$ *which has continuous first and second derivatives in R and equals zero on the boundary* Γ *of R, then* $\alpha(x, y) = 0$ *everywhere in R.*

Proof. Suppose the function $\alpha(x, y)$ is nonzero, say positive, at some point in R. Then $\alpha(x, y)$ is also positive in some circle

$$(x - x_0)^2 + (y - y_0)^2 \leqslant \varepsilon^2 \tag{22}$$

contained in R, with center (x_0, y_0) and radius ε. If we set $h(x, y) = 0$ outside the circle (22) and

$$h(x, y) = [(x - x_0)^2 + (y - y_0)^2 - \varepsilon^2]^3$$

inside the circle, then $h(x)$ satisfies the conditions of the lemma. However, in this case, (21) reduces to an integral over the circle (22) and is obviously positive. This contradiction proves the lemma.

In order to apply the necessary condition for an extremum of the functional (20), i.e., $\delta J = 0$, we must first calculate the variation δJ. Let $h(x, y)$ be an arbitrary function which has continuous first and second derivatives in the region R and vanishes on the boundary Γ of R. Then if $z(x, y)$ belongs to the domain of definition of the functional (20), so does $z(x, y) + h(x, y)$. Since

$$\Delta J = J[z + h] - J[z] = \iint_R [F(x, y, z + h, z_x + h_x, z_y + h_y)$$
$$- F(x, y, z, z_x, z_y)]\, dx\, dy,$$

it follows by using Taylor's theorem that

$$\Delta J = \iint_R (F_z h + F_{z_x} h_x + F_{z_y} h_y)\, dx\, dy + \cdots,$$

where the dots denote terms of order higher than 1 relative to h, h_x and h_y. The integral on the right represents the principal linear part of the increment ΔJ, and hence the variation of $J[z]$ is

$$\delta J = \iint_R (F_z h + F_{z_x} h_x + F_{z_y} h_y)\, dx\, dy.$$

Next, we observe that

$$\iint_R (F_{z_x} h_x + F_{z_y} h_y)\, dx\, dy$$
$$= \iint_R \left[\frac{\partial}{\partial x} (F_{z_x} h) + \frac{\partial}{\partial y} (F_{z_y} h) \right] dx\, dy - \iint_R \left(\frac{\partial}{\partial x} F_{z_x} + \frac{\partial}{\partial y} F_{z_y} \right) h\, dx\, dy$$
$$= \int_\Gamma (F_{z_x} h\, dy - F_{z_y} h\, dx) - \iint_R \left(\frac{\partial}{\partial x} F_{z_x} + \frac{\partial}{\partial y} F_{z_y} \right) h\, dx\, dy,$$

where in the last step we have used Green's theorem[11]

$$\iint_R \left(\frac{\partial Q}{\partial x} - \frac{\partial P}{\partial y} \right) dx\, dy = \int_\Gamma (P\, dx + Q\, dy).$$

The integral along Γ is zero, since $h(x, y)$ vanishes on Γ, and hence, comparing the last two formulas, we find that

$$\delta J = \iint_R \left(F_z - \frac{\partial}{\partial x} F_{z_x} - \frac{\partial}{\partial y} F_{z_y} \right) h(x, y)\, dx\, dy. \tag{23}$$

---

11 See e.g., D. V. Widder, *Advanced Calculus*, second edition, Dover, Mineola, N.Y. (1961), p. 223.

Thus, the condition $\delta J = 0$ implies that the double integral (23) vanishes for any $h(x, y)$ satisfying the stipulated conditions. According to the lemma, this leads to the following second-order *partial* differential equation, again known as *Euler's equation*:

$$F_z - \frac{\partial}{\partial x} F_{z_x} - \frac{\partial}{\partial y} F_{z_y} = 0. \tag{24}$$

We are looking for a solution of (24) which takes given values on the boundary Γ.

Example. *Find the surface of least area spanned by a given contour.*

This problem reduces to finding the minimum of the functional

$$J[z] = \iint_R \sqrt{1 + z_x^2 + z_y^2} \, dx \, dy,$$

so that Euler's equation has the form

$$r(1 + q^2) - 2spq + t(1 + p^2) = 0, \tag{25}$$

where

$$p = z_x, \quad q = z_y, \quad r = z_{xx}, \quad s = z_{xy}, \quad t = z_{yy}.$$

Equation (25) has a simple geometric meaning, which we explain by using the formula

$$M = \frac{1}{2}\left(\frac{1}{\varkappa_1} + \frac{1}{\varkappa_2}\right) = \frac{Eg - 2Ff + Ge}{2(EG - F^2)}$$

for the mean curvature of the surface, where E, F, G and e, f, g are the coefficients of the first and second fundamental quadratic forms of the surface.[12] If the surface is given by an explicit equation of the form $z = z(x, y)$, then

$$E = 1 + p^2, \quad F = pq, \quad G = 1 + q^2,$$

$$e = \frac{r}{\sqrt{1 + p^2 + q^2}}, \quad f = \frac{s}{\sqrt{1 + p^2 + q^2}}, \quad g = \frac{t}{\sqrt{1 + p^2 + q^2}},$$

and hence

$$M = \frac{(1 + p^2)t - 2spq + (1 + q^2)r}{\sqrt{1 + p^2 + q^2}}.$$

Here, the numerator coincides with the left-hand side of Euler's equation (25). Thus, (25) implies that the mean curvature of the required surface equals zero. Surfaces with zero mean curvature are called *minimal surfaces*.

---

[12] See e.g., D. V. Widder, *op. cit.*, Chap. 3, Sec. 6, and E. Kreysig, *Differential Geometry*, University of Toronto Press, Toronto (1959), Chap. 4. Here, $\varkappa_1$ and $\varkappa_2$ denote the principal normal curvatures of the surface.

6. A Simple Variable End Point Problem

There are, of course, many other kinds of variational problems besides the "simplest" variational problem considered so far, and such problems will be studied in Chapters 2 and 3. However, this is a suitable place for acquainting the reader with one of these problems, i.e., the *variable end point problem*, a particular case of which can be stated as follows: *Among all curves whose end points lie on two given vertical lines* $x = a$ *and* $x = b$, *find the curve for which the functional*

$$J[y] = \int_a^b F(x, y, y')\, dx \qquad (26)$$

has an extremum.[13]

We begin by calculating the variation δJ of the functional (26). As before, δJ means the principal linear part of the increment

$$\Delta J = J[y + h] - J[y] = \int_a^b [F(x, y + h, y' + h') - F(x, y, y')]\, dx.$$

Using Taylor's theorem to expand the integrand, we obtain

$$\Delta J = \int_a^b (F_y h + F_{y'} h')\, dx + \cdots,$$

where the dots denote terms of order higher than 1 relative to h and h', and hence

$$\delta J = \int_a^b (F_y h + F_{y'} h')\, dx.$$

Here, unlike the fixed end point problem, $h(x)$ need no longer vanish at the points a and b, so that integration by parts now gives[14]

$$
\begin{aligned}
\delta J &= \int_a^b \left(F_y - \frac{d}{dx} F_{y'}\right) h(x)\, dx + F_{y'} h(x)\big|_{x=a}^{x=b} \\
&= \int_a^b \left(F_y - \frac{d}{dx} F_{y'}\right) h(x)\, dx + F_{y'}\big|_{x=b}\, h(b) - F_{y'}\big|_{x=a}\, h(a).
\end{aligned} \qquad (27)
$$

We first consider functions $h(x)$ such that $h(a) = h(b) = 0$. Then, as in the simplest variational problem, the condition $\delta J = 0$ implies that

$$F_y - \frac{d}{dx} F_{y'} = 0. \qquad (28)$$

Therefore, in order for the curve $y = y(x)$ to be a solution of the variable end point problem, y must be an extremal, i.e., a solution of Euler's equation.

---

[13] The more general case where the end points lie on two given curves $y = \varphi(x)$ and $y = \psi(x)$ is treated in Sec. 14.

[14] As usual, $f(x)|_{x=a}^{x=b}$ stands for $f(b) - f(a)$.

But if y is an extremal, the integral in the expression (27) for δJ vanishes, and then the condition $\delta J = 0$ takes the form

$$F_{y'}|_{x=b}\, h(b) - F_{y'}|_{x=a}\, h(a) = 0,$$

from which it follows that

$$F_{y'}|_{x=a} = 0, \qquad F_{y'}|_{x=b} = 0, \tag{29}$$

since $h(x)$ is arbitrary.   Thus, to solve the variable end point problem, we must first find a general integral of Euler's equation (28), and then use the conditions (29), sometimes called the *natural boundary conditions*, to determine the values of the arbitrary constants.

Besides the case of fixed end points and the case of variable end points, we can also consider the *mixed case*, where one end is fixed and the other is variable.   For example, suppose we are looking for an extremum of the functional (26) with respect to the class of curves joining a given point A (with abscissa a) and an arbitrary point of the line $x = b$.   In this case, the conditions (29) reduce to the single condition

$$F_{y'}|_{x=b} = 0,$$

and $y(a) = A$ serves as the second boundary condition.

Example.   *Starting from the point* $P = (a, A)$, *a heavy particle slides down a curve in the vertical plane. Find the curve such that the particle reaches the vertical line* $x = b$ ($\neq a$) *in the shortest time.*   (This is a variant of the brachistochrone problem, p. 3.)

For simplicity, we assume that the original point coincides with the origin of coordinates.   Since the velocity of motion along the curve equals

$$v = \frac{ds}{dt} = \sqrt{1 + y'^2}\, \frac{dx}{dt},$$

we have

$$dt = \frac{\sqrt{1 + y'^2}}{v}\, dx = \frac{\sqrt{1 + y'^2}}{\sqrt{2gy}}\, dx,$$

so that the transit time T is given by the equation

$$T = \int \frac{\sqrt{1 + y'^2}}{\sqrt{2gy}}\, dx.$$

The general solution of the corresponding Euler equation consists of a family of cycloids

$$x = r(\theta - \sin \theta) + c, \qquad y = r(1 - \cos \theta).$$

Since the curve must pass through the origin, we must have $c = 0$. To determine r, we use the second condition

$$F_{y'} = \frac{y'}{\sqrt{2gy}\,\sqrt{1 + y'^2}} = 0 \qquad \text{for} \quad x = b,$$

i.e., $y' = 0$ for $x = b$, which means that the tangent to the curve at its right end point must be horizontal. It follows that $r = b/\pi$, and hence the required curve is given by the equations

$$x = \frac{b}{\pi}(\theta - \sin\theta), \qquad y = \frac{b}{\pi}(1 - \cos\theta).$$

7. The Variational Derivative

In Sec. 3.2 we introduced the concept of the differential of a functional. We now introduce the concept of the *variational* (or *functional*) *derivative*, which plays the same role for functionals as the concept of the partial derivative plays for functions of n variables. We begin by considering functionals of the type

$$J[y] = \int_a^b F(x, y, y')\,dx, \qquad y(a) = A, \quad y(b) = B, \qquad (30)$$

corresponding to the simplest variational problem. Our approach is to first go from the variational problem to an n-dimensional problem, and then pass to the limit $n \to \infty$.

Thus, we divide the interval $[a, b]$ into $n + 1$ equal subintervals by introducing the points

$$x_0 = a, \ x_1, \ldots, x_n, \ x_{n+1} = b, \qquad (x_{i+1} - x_i = \Delta x),$$

and we replace the smooth function $y(x)$ by the polygonal line with vertices

$$(x_0, y_0), (x_1, y_1), \ldots, (x_n, y_n), (x_{n+1}, y_{n+1}),$$

where $y_i = y(x_i)$.[15] Then (30) can be approximated by the sum

$$J(y_1, \ldots, y_n) \equiv \sum_{i=0}^{n} F\left(x_i, y_i, \frac{y_{i+1} - y_i}{\Delta x}\right)\Delta x, \qquad (31)$$

which is a function of n variables. (Recall that $y_0 = A$ and $y_{n+1} = B$ are fixed.)

Next, we calculate the partial derivatives

$$\frac{\partial J(y_1, \ldots, y_n)}{\partial y_k},$$

and we consider what happens to these derivatives as the number of points of subdivision increases without limit. Observing that each variable y_k

---

[15] This is the *method of finite differences* (cf. Secs. 1, 40).

in (31) appears in just two terms, corresponding to $i = k$ and $i = k - 1$, we find that

$$\frac{\partial J}{\partial y_k} = F_y\left(x_k, y_k, \frac{y_{k+1} - y_k}{\Delta x}\right)\Delta x$$
$$+ F_{y'}\left(x_{k-1}, y_{k-1}, \frac{y_k - y_{k-1}}{\Delta x}\right) - F_{y'}\left(x_k, y_k, \frac{y_{k+1} - y_k}{\Delta x}\right). \tag{32}$$

As $\Delta x \to 0$, i.e., as the number of points of subdivision increases without limit, the right-hand side of (32) obviously goes to zero, since it is a quantity of order Δx. In order to obtain a limit which is in general nonzero as $\Delta x \to 0$, we divide (32) by Δx, obtaining

$$\frac{\partial J}{\partial y_k \Delta x} = F_y\left(x_k, y_k, \frac{y_{k+1} - y_k}{\Delta x}\right)$$
$$- \frac{1}{\Delta x}\left[F_{y'}\left(x_k, y_k, \frac{y_{k+1} - y_k}{\Delta x}\right) - F_{y'}\left(x_{k-1}, y_{k-1}, \frac{y_k - y_{k-1}}{\Delta x}\right)\right]. \tag{33}$$

We note that the expression $\partial y_k \Delta x$ appearing in the denominator on the left has a direct geometric meaning, and is in fact just the area of the region lying between the solid and the dashed curves in Figure 3.

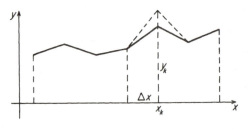

FIGURE 3

As $\Delta x \to 0$, the expression (33) converges to the limit

$$\frac{\delta J}{\delta y} \equiv F_y(x, y, y') - \frac{d}{dx} F_{y'}(x, y, y'),$$

called the *variational derivative* of the functional (30). We see that the variational derivative $\delta J/\delta y$ is just the left-hand side of Euler's equation (28), and hence the meaning of Euler's equation is just that the variational derivative of the functional under consideration should vanish at every point. This is the analog of the situation encountered in elementary analysis, where a necessary condition for a function of n variables to have an extremum is that all its partial derivatives vanish.

In the general case, the variational derivative is defined as follows: Let $J[y]$ be a functional depending on the function $y(x)$, and suppose we give $y(x)$ an increment $h(x)$ which is different from zero only in the neighborhood of a point x_0. Dividing the corresponding increment $J[y + h] - J[y]$ of the functional by the area $\Delta\sigma$ lying between the curve $y = h(x)$ and the x-axis,[16] we obtain the ratio

$$\frac{J[y + h] - J[y]}{\Delta\sigma}. \tag{34}$$

Next, we let the area $\Delta\sigma$ go to zero in such a way that both max $|h(x)|$ and the length of the interval in which $h(x)$ is nonvanishing go to zero. Then, if the ratio (34) converges to a limit as $\Delta\sigma \to 0$, this limit is called the *variational derivative* of the functional $J[y]$ at the point x_0 [for the curve $y = y(x)$], and is denoted by

$$\left.\frac{\delta J}{\delta y}\right|_{x=x_0}$$

It can be shown that the analogs of all the familiar rules obeyed by ordinary derivatives (e.g., the formulas for differentiating sums and products of functions, composite functions, etc.) are valid for variational derivatives.

Remark. It is clear from the definition of the variational derivative that if $h(x)$ is different from zero in a neighborhood of the point x_0 and if $\Delta\sigma$ is the area between the curve $y = h(x)$ and the x-axis, then

$$\Delta J \equiv J[y + h] - J[y] = \left\{\left.\frac{\delta J}{\delta y}\right|_{x=x_0} + \varepsilon\right\}\Delta\sigma,$$

where $\varepsilon \to 0$ as both max $|h(x)|$ and the length of the interval in which $h(x)$ is nonvanishing go to zero. It follows that in terms of the variational derivative, the differential or variation of the functional $J[y]$ at the point x_0 [for the curve $y = y(x)$] is given by the formula

$$\delta J = \left.\frac{\delta J}{\delta y}\right|_{x=x_0}\Delta\sigma.$$

8. Invariance of Euler's Equation

Suppose that instead of the rectangular plane coordinates x and y, we introduce curvilinear coordinates u and v, where

$$\begin{aligned}x &= x(u, v), \\ y &= y(u, v),\end{aligned} \qquad \begin{vmatrix} x_u & x_v \\ y_u & y_v \end{vmatrix} \neq 0. \tag{35}$$

Then the curve given by the equation $y = y(x)$ in the xy-plane corresponds to the curve given by some equation

$$v = v(u)$$

---

[16] $\Delta\sigma$ can also be regarded as the area between the curves $y = y(x)$ and $y = y(x) + h(x)$.

in the uv-plane. When we make the change of variables (35), the functional

$$J[y] = \int_a^b F(x, y, y')\, dx$$

goes into the functional

$$J_1[v] = \int_{a_1}^{b_1} F\left[x(u, v),\, y(u, v),\, \frac{y_u + y_v v'}{x_u + x_v v'}\right](x_u + x_v v')\, du$$

$$= \int_{a_1}^{b_1} F_1(u, v, v')\, du,$$

where

$$F_1(u, v, v') = F\left[x(u, v),\, y(u, v),\, \frac{y_u + y_v v'}{x_u + x_v v'}\right](x_u + x_v v').$$

We now show that if $y = y(x)$ satisfies the Euler equation

$$\frac{\partial F}{\partial y} - \frac{d}{dx}\frac{\partial F}{\partial y'} = 0 \qquad (36)$$

corresponding to the original functional $J[y]$, then $v = v(u)$ satisfies the Euler equation

$$\frac{\partial F_1}{\partial v} - \frac{d}{du}\frac{\partial F_1}{\partial v'} = 0 \qquad (37)$$

corresponding to the new functional $J_1[v]$. To prove this, we use the concept of the variational derivative, introduced in the preceding section. Let $\Delta\sigma$ denote the area bounded by the curves $y = y(x)$ and $y = y(x) + h(x)$, and let $\Delta\sigma_1$ denote the area bounded by the corresponding curves $v = v(u)$ and $v = v(u) + \eta(u)$ in the uv-plane. By the standard formula for the transformation of areas, the limit as $\Delta\sigma,\, \Delta\sigma_1 \to 0$ of the ratio $\Delta\sigma/\Delta\sigma_1$ approaches the Jacobian

$$\begin{vmatrix} x_u & x_v \\ y_u & y_v \end{vmatrix},$$

which by hypothesis is nonzero. Thus, if

$$\lim_{\Delta\sigma \to 0} \frac{J[y + h] - J[y]}{\Delta\sigma} = 0,$$

then

$$\lim_{\Delta\sigma_1 \to 0} \frac{J_1[v + \eta] - J_1[v]}{\Delta\sigma_1} = 0$$

as well. It follows that $v(u)$ satisfies (37) if $y(x)$ satisfies (36). In other words, whether or not a curve is an extremal is a property which is independent of the choice of the coordinate system.

In solving Euler's equation, changes of variables can often be used to advantage. Because of the invariance property just proved, the change of variables can be made directly in the integral representing the functional rather than in Euler's equation, and we can then write Euler's equation for the new integral.

Example. Suppose we are looking for the extremals of the functional

$$J[r] = \int_{\varphi_0}^{\varphi_1} \sqrt{r^2 + r'^2}\, d\varphi, \qquad (38)$$

where $r = r(\varphi)$. The corresponding Euler equation has the form

$$\frac{r}{\sqrt{r^2 + r'^2}} - \frac{d}{d\varphi} \frac{r'}{\sqrt{r^2 + r'^2}} = 0.$$

The change of variables

$$x = r \cos \varphi, \qquad y = r \sin \varphi$$

transforms (38) into an integral of the form

$$\int_{x_0}^{x_1} \sqrt{1 + y'^2}\, dx,$$

which has the Euler equation

$$y'' = 0,$$

with general solution

$$y = \alpha x + \beta.$$

Therefore, the solution of (38) is

$$r \sin \varphi = \alpha r \cos \varphi + \beta.$$

PROBLEMS

1. Use the method of finite differences (Sec. 1) to find the shortest plane curve joining two points A and B.

2. A set $\mathcal{M}$ in a normed linear space $\mathcal{R}$ is said to be *convex* if $\mathcal{M}$ contains all elements of the form $\alpha x + \beta y$, where $\alpha, \beta \geqslant 0$, $\alpha + \beta = 1$, provided that $\mathcal{M}$ contains x and y. Prove that the set of all elements $x \in \mathcal{R}$ satisfying the inequality $\|x - x_0\| \leqslant c$, where x_0 is a fixed element of $\mathcal{R}$ and $c > 0$, is convex.

3. Show that the set $\overline{\mathscr{C}}(a, b)$ of all continuous functions defined on the interval $[a, b]$, equipped with the norm

$$\|y\| = \left\{ \int_a^b |y(x)|^2\, dx \right\}^{1/2},$$

forms a normed linear space.

4. An infinite sequence of elements $y_1, y_2, \ldots$ of elements of a normed linear space $\mathcal{R}$ is called a *Cauchy sequence* (or *fundamental sequence*) if, given any $\varepsilon > 0$, there exists an integer $N = N(\varepsilon)$ such that $\|y_m - y_n\| < \varepsilon$, provided that $m > N$, $n > N$. A normed linear space $\mathcal{R}$ is said to be complete if every Cauchy sequence in $\mathcal{R}$ converges to some element in $\mathcal{R}$. Prove that the space $\overline{\mathscr{C}}(a, b)$ introduced in the preceding problem is not complete, but that the space $\mathscr{C}(a, b)$ introduced in Sec. 2 is complete.

Comment. See e.g., G. E. Shilov, *op. cit.*, p. 249.

5. Prove that any norm defined on a linear space $\mathscr{R}$ is a continuous functional on $\mathscr{R}$.

6. Suppose the norm of the space $\mathscr{D}_n(a, b)$ is defined as

$$\|y\| = \max_{a \leqslant x \leqslant b} \{|y(x)|, |y'(x)|, \ldots, |y^{(n)}(x)|\},$$

instead of

$$\|y\| = \sum_{i=0}^{n} \max_{a \leqslant x \leqslant b} |y^{(i)}(x)|,$$

as on p. 7. Prove that any functional on $\mathscr{D}_n(a, b)$ which is continuous with respect to one of these norms is continuous with respect to the other.

7. Let $J[y]$ be the arc-length functional, defined for all $y \in \mathscr{D}_1(a, b)$. Show that $J[y]$ is lower semicontinuous with respect to the norm of the space $\mathscr{C}(a, b)$.

Comment. As remarked in footnote 5, p. 8, $J[y]$ is not continuous with respect to the norm of $\mathscr{C}(a, b)$.

8. Let $\varphi[h]$ be a linear functional defined on a normed linear space $\mathscr{R}$. Prove that if $\varphi[h]$ is continuous for $h = 0$, it is continuous for all $h \in \mathscr{R}$.

9. Prove that a linear functional $\varphi[h]$ cannot have an extremum unless $\varphi[h] \equiv 0$.

10. Prove that if two linear functionals $\varphi[h]$ and $\psi[h]$ defined on the same space vanish on the same set of elements, then $\varphi[h] = \lambda \psi[h]$, where λ is a constant.

11. Show that constants c_0 and c_1 can always be chosen satisfying the conditions (7) used to prove Lemma 3, p. 10.

12. Prove that the square of a differentiable functional is differentiable, and write a formula for its differential (variation).

13. Prove that if two differentiable functionals defined on the same normed linear space have the same differential at every point of the space, then they differ by a constant.

14. Analyze the variational problems corresponding to the following functionals, where in each case $y(0) = 0$, $y(1) = 1$:

a) $\int_0^1 y' \, dx$; b) $\int_0^1 yy' \, dx$; c) $\int_0^1 xyy' \, dx$.

15. Find the extremals of the following functionals:

a) $\int_a^b (y^2 + y'^2 - 2y \sin x) \, dx$; b) $\int_a^b \frac{y'^2}{x^3} \, dx$;

c) $\int_a^b (y^2 - y'^2 - 2y \cosh x) \, dx$; d) $\int_a^b (y^2 + y'^2 + 2ye^x) \, dx$;

e) $\int_a^b (y^2 - y'^2 - 2y \sin x) \, dx$.

Ans. b) $y = C_1 x^4 + C_2$; d) $y = \frac{1}{2} x e^x + C_1 e^x + C_2 e^{-x}$.

16. Prove the uniqueness part of Bernstein's theorem (p. 16).

Hint. Let $\Delta(x) = \varphi_2(x) - \varphi_1(x)$, where $\varphi_1(x)$ and $\varphi_2(x)$ are two solutions of (15), write an expression for $\Delta''(x)$ and use the condition $F_y(x, y, y') > k$.

17. Prove that one and only one extremal of each of the functionals

$$\int e^{-2v^2}(y'^2 - 1)\, dx, \qquad \int (y^2 + y'\tan^{-1} y' - \ln \sqrt{1 + y'^2})\, dx$$

passes through any two points of the plane with different abscissas.

Hint. Apply Bernstein's theorem.

18. Find the general solution of the Euler equation corresponding to the functional

$$J[y] = \int_a^b f(x)\sqrt{1 + y'^2}\, dx,$$

and investigate the special cases $f(x) = \sqrt{x}$ and $f(x) = x$.

Comment. The case $f(x) = 1/x$ is treated in Example 1, p. 19.

19. Find all minimal surfaces whose equations have the form $z = \varphi(x) + \psi(y)$.

Ans. $z = Ax + By + C$, $\qquad e^{a(z - z_0)} = \dfrac{\cos a(y - y_0)}{\cos a(x - x_0)}.$

20. Which curve minimizes the integral

$$\int_0^1 (\tfrac{1}{2}y'^2 + yy' + y' + y)\, dx,$$

when the values of y are not specified at the end points?

Ans. $y = \tfrac{1}{2}(x^2 - 3x + 1).$

21. Calculate the variational derivative at the point x_0 of the quadratic functional

$$J[y] = \int_a^b \int_a^b K(s, t)\, y(s)\, y(t)\, ds\, dt.$$

22. Find the extremals of the functional

$$\int \sqrt{x^2 + y^2}\, \sqrt{1 + y'^2}\, dx.$$

Hint. Use polar coordinates.

Ans. $x^2 \cos \alpha + 2xy \sin \alpha - y^2 \cos \alpha = \beta$, where α and β are constants.

2

FURTHER GENERALIZATIONS

In this chapter, we consider some further generalizations of the simplest variational problem. These include variational problems in spaces of dimension greater than two (Sec. 9), problems in parametric form (Sec. 10), problems involving higher derivatives (Sec. 11), and problems with subsidiary conditions (Sec. 12).

9. The Fixed End Point Problem for n Unknown Functions

Let $F(x, y_1, \ldots, y_n, z_1, \ldots, z_n)$ be a function with continuous first and second (partial) derivatives with respect to all its arguments. Consider the problem of finding necessary conditions for an extremum of a functional of the form

$$J[y_1, \ldots, y_n] = \int_a^b F(x, y_1, \ldots, y_n, y_1', \ldots, y_n') \, dx, \tag{1}$$

which depends on n continuously differentiable functions $y_1(x), \ldots, y_n(x)$ satisfying the boundary conditions

$$y_i(a) = A_i, \quad y_i(b) = B_i \qquad (i = 1, \ldots, n). \tag{2}$$

In other words, we are looking for an extremum of the functional (1) defined on the set of the set of smooth curves joining two fixed points in $(n + 1)$-dimensional Euclidean space $\mathscr{E}_{n+1}$. The problem of finding *geodesics*, i.e., shortest curves joining two points of some manifold, is of this type. The same kind of problem arises in geometric optics, in finding the paths along which light rays propagate in an inhomogeneous medium. In fact, according to *Fermat's principle*, light goes from a point P_0 to a point P_1 along the path for which the transit time is the smallest.

To find necessary conditions for the functional (1) to have an extremum, we first calculate its variation. Suppose we replace each $y_i(x)$ by a "varied" function $y_i(x) + h_i(x)$. By the *variation* δJ of the functional $J[y_1, \ldots, y_n]$, we mean the expression which is linear in h_i, h_i' $(i = 1, \ldots, n)$ and differs from the increment

$$\Delta J = J[y_1 + h_1, \ldots, y_n + h_n] - J[y_1, \ldots, y_n]$$

by a quantity of order higher than 1 relative to h_i, h_i' $(i = 1, \ldots, n)$. Since both $y_i(x)$ and $y_i(x) + h_i(x)$ satisfy the boundary conditions (2), for each i, it is clear that

$$h_i(a) = h_i(b) = 0 \qquad (i = 1, \ldots, n).$$

We now use Taylor's theorem, obtaining

$$\Delta J = \int_a^b [F(x, \ldots, y_i + h_i, y_i' + h_i', \ldots)\, dx - F(x, \ldots, y_i, y_i', \ldots)]\, dx$$

$$= \int_a^b \sum_{i=1}^n (F_{y_i} h_i + F_{y_i'} h_i')\, dx + \cdots,$$

where the dots denote terms of order higher than 1 relative to h_i, h_i' $(i = 1, \ldots, n)$. The last integral on the right represents the principal linear part of the increment ΔJ, and hence the variation of $J[y_1, \ldots, y_n]$ is

$$\delta J = \int_a^b \sum_{i=1}^n (F_{y_i} h_i + F_{y_i'} h_i')\, dx.$$

Since all the increments $h_i(x)$ are independent, we can choose one of them quite arbitrarily (as long as the boundary conditions are satisfied), setting all the others equal to zero. Therefore, the necessary condition $\delta J = 0$ for an extremum implies

$$\int_a^b (F_{y_i} h_i + F_{y_i'} h_i')\, dx = 0 \qquad (i = 1, \ldots, n).$$

Using Lemma 4 of Sec. 3.1, we obtain the following system of *Euler equations*:

$$F_{y_i} - \frac{d}{dx} F_{y_i'} = 0 \qquad (i = 1, \ldots, n). \tag{3}$$

Since (3) is a system of n second-order differential equations, its general solution contains $2n$ arbitrary constants, which are determined from the boundary conditions (2). Thus, we have proved the following

THEOREM. *A necessary condition for the curve*

$$y_i = y_i(x) \qquad (i = 1, \ldots, n)$$

to be an extremal of the functional

$$\int_a^b F(x, y_1, \ldots, y_n, y_1', \ldots, y_n')\, dx$$

is that the functions $y_i(x)$ satisfy the Euler equations (3).

Remark 1. We have just shown how to find a well-defined system of Euler equations (3) for every functional of the form (1). However, two different integrands F can lead to the same set of Euler equations. In fact, let

$$\Phi = \Phi(x, y_1, \ldots, y_n)$$

be any twice differentiable function, and let

$$\Psi(x, y_1, \ldots, y_n, y_1', \ldots, y_n') = \frac{\partial \Phi}{\partial x} + \sum_{i=1}^{n} \frac{\partial \Phi}{\partial y_i} y_i'. \tag{4}$$

Then we find at once by direct calculation that

$$\frac{\partial}{\partial y_i} - \frac{d}{dx}\left(\frac{\partial \Psi}{\partial y_i'}\right) \equiv 0,$$

and hence the functionals

$$\int_a^b F(x, y_1, \ldots, y_n, y_1', \ldots, y_n') \, dx \tag{5}$$

and

$$\int_a^b [F(x, y_1, \ldots, y_n, y_1', \ldots, y_n') + \Psi(x, y_1, \ldots, y_n, y_1', \ldots, y_n')] \, dx \tag{6}$$

lead to the same system of Euler equations.

Given any curve $y_i = y_i(x)$, the function (4) is just the derivative

$$\frac{d}{dx} \Phi[x, y_1(x), \ldots, y_n(x)].$$

Therefore, the integral

$$\int_a^b \Psi(x, y_1, \ldots, y_n, y_1', \ldots, y_n') \, dx = \int_a^b \frac{d\Phi}{dx} \, dx$$

takes the same value along all curves satisfying the boundary conditions (2). In other words, the functionals (5) and (6), defined on the class of functions satisfying (2), differ only by a constant. In particular, we can choose Φ in such a way that this constant vanishes (but $\Psi \not\equiv 0$).

Remark 2. Two functionals are said to be *equivalent* if they have the same extremals. According to Remark 1, two functionals of the form (1) are equivalent if their integrands differ by a function of the form (4). It is also clear that two functionals of this form are equivalent if their integrands differ by a constant factor $c \neq 0$. More generally, the functional (5) is equivalent to the functional (6) with F replaced by cF.

Example 1. *Propagation of light in an inhomogeneous medium.* Suppose that three-dimensional space is filled with an optically inhomogeneous medium, such that the velocity of propagation of light at each point is some function $v(x, y, z)$ of the coordinates of the point. According to Fermat's principle (see p. 34), light goes from one point to another along the curve for which the transit time of the light is the smallest.   If the curve joining two points A and B is specified by the equations

$$y = y(x), \qquad z = z(x),$$

the time it takes light to traverse the curve equals

$$\int_a^b \frac{\sqrt{1 + y'^2 + z'^2}}{v(x, y, z)} \, dx.$$

Writing the system of Euler equations for this functional, i.e.,

$$\frac{\partial v}{\partial y} \frac{\sqrt{1 + y'^2 + z'^2}}{v^2} + \frac{d}{dx} \frac{y'}{v\sqrt{1 + y'^2 + z'^2}} = 0,$$

$$\frac{\partial v}{\partial z} \frac{\sqrt{1 + y'^2 + z'^2}}{v^2} + \frac{d}{dx} \frac{z'}{v\sqrt{1 + y'^2 + z'^2}} = 0,$$

we obtain the differential equations for the curves along which the light propagates.

Example 2. Geodesics.   Suppose we have a surface σ specified by a vector equation[1]

$$\mathbf{r} = \mathbf{r}(u, v). \tag{7}$$

The shortest curve lying on σ and connecting two points of σ is called the *geodesic* connecting the two points.   Clearly, the equations for the geodesics of σ are the Euler equations of the corresponding variational problem, i.e., the problem of finding the minimum distance (measured along σ) between two points of σ.

A curve lying on the surface (7) can be specified by the equations

$$u = u(t), \qquad v = v(t).$$

The arc length between the points corresponding to the values t_1 and t_2 of the parameter t equals

$$J[u, v] = \int_{t_0}^{t_1} \sqrt{Eu'^2 + 2Fu'v' + Gv'^2} \, dt, \tag{8}$$

where E, F and G are the coefficients of the first fundamental (quadratic) form of the surface (7), i.e.,[2]

$$E = \mathbf{r}_u \cdot \mathbf{r}_u, \qquad F = \mathbf{r}_u \cdot \mathbf{r}_v, \qquad G = \mathbf{r}_v \cdot \mathbf{r}_v.$$

---

[1] Here, vectors are indicated by boldface letters, and $\mathbf{a} \cdot \mathbf{b}$ denotes the scalar product of the vectors $\mathbf{a}$ and $\mathbf{b}$.

[2] See D. V. Widder, *op. cit.*, p. 110.

Writing the Euler equations for the functional (8), we obtain

$$\frac{E_u u'^2 + 2F_u u'v' + G_u v'^2}{\sqrt{Eu'^2 + 2Fu'v' + Gv'^2}} - \frac{d}{dt} \frac{2(Eu' + Fv')}{\sqrt{Eu'^2 + 2Fu'v' + Gv'^2}} = 0,$$

$$\frac{E_v u'^2 + 2F_v u'v' + G_v v'^2}{\sqrt{Eu'^2 + 2Fu'v' + Gv'^2}} - \frac{d}{dt} \frac{2(Fu' + Gv')}{\sqrt{Eu'^2 + 2Fu'v' + Gv'^2}} = 0.$$

As a very simple illustration of these considerations, we now find the geodesics of the circular cylinder

$$\mathbf{r} = (a \cos \varphi, a \sin \varphi, z), \tag{9}$$

where the variables φ and z play the role of the parameters u and v. Since the coefficients of the first fundamental form of the cylinder (9) are

$$E = a^2, \qquad F = 0, \qquad G = 1,$$

the geodesics of the cylinder have the equations

$$\frac{d}{dt} \frac{a^2 \varphi'}{\sqrt{a^2 \varphi'^2 + z'^2}} = 0, \qquad \frac{d}{dt} \frac{z'}{\sqrt{a^2 \varphi'^2 + z'^2}} = 0,$$

i.e.,

$$\frac{a^2 \varphi'}{\sqrt{a^2 \varphi'^2 + z'^2}} = C_1, \qquad \frac{z'}{\sqrt{a^2 \varphi'^2 + z'^2}} = C_2.$$

Dividing the second of these equations by the first, we obtain

$$\frac{dz}{d\varphi} = c_1,$$

which has the solution

$$z = c_1 \varphi + c_2,$$

representing a two-parameter family of helical lines lying on the cylinder (9).

The concept of a geodesic can be defined not only for surfaces, but also for higher-dimensional manifolds. Clearly, finding the geodesics of an n-dimensional manifold reduces to solving a variational problem for a functional depending on n functions.

10. Variational Problems in Parametric Form

So far, we have considered functionals of curves given by explicit equations, e.g., by equations of the form

$$y = y(x) \tag{10}$$

in the two-dimensional case. However, it is often more convenient to consider functionals of curves given in parametric form, and in fact we have already encountered this case in Example 2 of the preceding section (involving geodesics on a surface). Moreover, in problems involving closed curves (like the isoperimetric problem mentioned on p. 3), it is usually impossible to get along without representing the curves in parametric form. Thus, in this section, we extend our previous results to the case where the curves are given parametrically, confining ourselves to the simplest variational problem.

Suppose that in the functional

$$\int_{x_0}^{x_1} F(x, y, y') \, dx, \tag{11}$$

we wish to regard the argument y as a curve which is given in parametric form, rather than in the form (10). Then (11) can be written as

$$\int_{t_0}^{t_1} F\left[x(t), y(t), \frac{\dot{y}(t)}{\dot{x}(t)}\right] \dot{x}(t) \, dt = \int_{t_0}^{t_1} \Phi(x, y, \dot{x}, \dot{y}) \, dt \tag{12}$$

(where the overdot denotes differentiation with respect to t), i.e., as a functional depending on two unknown functions $x(t)$ and $y(t)$. The function Φ appearing in the right-hand side of (12) does not involve t explicitly, and is *positive-homogeneous of degree 1* in $\dot{x}(t)$ and $\dot{y}(t)$, which means that

$$\Phi(x, y, \lambda\dot{x}, \lambda\dot{y}) \equiv \lambda\Phi(x, y, \dot{x}, \dot{y}) \tag{13}$$

for every $\lambda > 0$.[3]

Conversely, let

$$\int_{t_0}^{t_1} \Phi(x, y, \dot{x}, \dot{y}) \, dt$$

be a functional whose integrand Φ does not involve t explicitly and is positive-homogeneous of degree 1 in $\dot{x}$ and $\dot{y}$. We now show that the value of such a functional depends only on the curve in the xy-plane defined by the parametric equations $x = x(t)$, $y = y(t)$, and not on the functions $x(t)$, $y(t)$ themselves, i.e., that if we go from t to some new parameter τ by setting

$$t = t(\tau),$$

where $dt/d\tau > 0$ and the interval $[t_0, t_1]$ goes into $[\tau_0, \tau_1]$, then

$$\int_{\tau_0}^{\tau_1} \Phi\left(x, y, \frac{dx}{d\tau}, \frac{dy}{d\tau}\right) d\tau = \int_{t_0}^{t_1} \Phi(x, y, \dot{x}, \dot{y}) \, dt.$$

---

[3] The example of the arc-length functional

$$\int_{t_0}^{t_1} \sqrt{\dot{x}^2 + \dot{y}^2} \, dt,$$

whose value does not depend on the direction in which the curve $x = x(t)$, $y = y(t)$ is traversed, shows why (13) does not hold for $\lambda < 0$.

In fact, since Φ is positive-homogeneous of degree 1 in $\dot{x}$ and $\dot{y}$, it follows that

$$\int_{\tau_0}^{\tau_1} \Phi\left(x, y, \frac{dx}{d\tau}, \frac{dy}{d\tau}\right) d\tau = \int_{\tau_0}^{\tau_1} \Phi\left(x, y, \dot{x}\frac{dt}{d\tau}, \dot{y}\frac{dt}{d\tau}\right) d\tau$$

$$= \int_{\tau_0}^{\tau_1} \Phi(x, y, \dot{x}, \dot{y}) \frac{dt}{d\tau} d\tau = \int_{t_0}^{t_1} \Phi(x, y, \dot{x}, \dot{y}) \, dt,$$

as asserted. Thus, we have proved the following

THEOREM. *A necessary and sufficient condition for the functional*

$$\int_{t_0}^{t_1} \Phi(t, x, y, \dot{x}, \dot{y}) \, dt$$

to depend only on the curve in the xy-plane defined by the parametric equations $x = x(t)$, $y = y(t)$ and not on the choice of the parametric representation of the curve, is that the integrand Φ should not involve t explicitly and should be a positive-homogeneous function of degree 1 in $\dot{x}$ and $\dot{y}$.

Now, suppose some parameterization of the curve $y = y(x)$ reduces the functional (11) to the form

$$\int_{t_0}^{t_1} F\left(x, y, \frac{\dot{y}}{\dot{x}}\right) \dot{x} \, dt = \int_{t_0}^{t_1} \Phi(x, y, \dot{x}, \dot{y}) \, dt. \tag{14}$$

The variational problem for the right-hand side of (14) leads to the pair of Euler equations

$$\Phi_x - \frac{d}{dt}\Phi_{\dot{x}} = 0, \qquad \Phi_y - \frac{d}{dt}\Phi_{\dot{y}} = 0, \tag{15}$$

which must be equivalent to the single Euler equation

$$F_y - \frac{d}{dx}F_{y'} = 0,$$

corresponding to the variational problem for the original functional (11). Hence, the equations (15) cannot be independent, and in fact it is easily verified that they are connected by the identity

$$\dot{x}\left(\Phi_x - \frac{d}{dt}\Phi_{\dot{x}}\right) + \dot{y}\left(\Phi_y - \frac{d}{dt}\Phi_{\dot{y}}\right) = 0. \tag{16}$$

We shall discuss this point further in Sec. 37.5.

11. Functionals Depending on Higher-Order Derivatives

So far, we have considered functionals of the form

$$\int_a^b F(x, y, y') \, dx,$$

depending on the function $y(x)$ and its first derivative $y'(x)$, or of the more general form

$$\int_a^b F(x, y_1, \ldots, y_n, y_1', \ldots, y_n') \, dx,$$

depending on several functions $y_i(x)$ and their first derivatives $y_i'(x)$. However, many problems (e.g., in the theory of elasticity) involve functionals whose integrands contain not only $y_i(x)$ and $y_i'(x)$, but also higher-order derivatives $y_i''(x), y_i'''(x), \ldots$ The method given above for finding extrema of functionals (in the context of necessary conditions for weak extrema) can be carried over to this more general case without essential changes. For simplicity, we confine ourselves to the case of a single unknown function $y(x)$.

Thus, let $F(x, y, z_1, \ldots, z_n)$ be a function with continuous first and second (partial) derivatives with respect to all its arguments, and consider a functional of the form

$$J[y] = \int_a^b F(x, y, y', \ldots, y^{(n)}) \, dx. \tag{17}$$

Then we pose the following problem: *Among all functions $y(x)$ belonging to the space $\mathscr{D}_n(a, b)$ and satisfying the conditions*

$$\begin{aligned} y(a) &= A_0, \ y'(a) = A_1, \ \ldots, \ y^{(n-1)}(a) = A_{n-1}, \\ y(b) &= B_0, \ y'(b) = B_1, \ \ldots, \ y^{(n-1)}(b) = B_{n-1}, \end{aligned} \tag{18}$$

find the function for which (17) *has an extremum.* To solve this problem, we start from the general result which states that a necessary condition for a functional $J[y]$ to have an extremum is that its variation vanish (Theorem 2, p. 13). Thus, suppose we replace $y(x)$ by the "varied" function $y(x) + h(x)$, where $h(x)$, like $y(x)$, belongs to $\mathscr{D}_n(a, b)$.[4] By the *variation* δJ of the functional $J[y]$, we mean the expression which is linear in $h, h', \ldots, h^{(n)}$, and which differs from the increment

$$\Delta J = J[y + h] - J[y]$$

by a quantity of order higher than 1 relative to $h, h', \ldots, h^{(n)}$. Since both $y(x)$ and $y(x) + h(x)$ satisfy the boundary conditions (18), it is clear that

$$\begin{aligned} h(a) &= h'(a) = \cdots = h^{(n-1)}(a) = 0, \\ h(b) &= h'(b) = \cdots = h^{(n-1)}(b) = 0. \end{aligned} \tag{19}$$

Next, we use Taylor's theorem, obtaining

$$\begin{aligned} \Delta J &= \int_a^b [F(x, y + h, y' + h', \ldots, y^{(n)} + h^{(n)}) - F(x, y, y', \ldots, y^{(n)})] \, dx \\ &= \int_a^b (F_y h + F_{y'} h' + \cdots F_{y^{(n)}} h^{(n)}) \, dx + \cdots, \end{aligned}$$

---

[4] The increment $h(x)$ is often called the *variation of $y(x)$*. In problems involving "fixed end point conditions" like (18), we often write $h(x) = \delta y(x)$.

where the dots denote terms of order higher than 1 relative to $h, h', \ldots, h^{(n)}$. The last integral on the right represents the principal linear part of the increment ΔJ, and hence the variation of $J[y]$ is

$$\delta J = \int_a^b (F_y h + F_{y'} h' + \cdots + F_{y^{(n)}} h^{(n)}) \, dx.$$

Therefore, the necessary condition $\delta J = 0$ for an extremum implies that

$$\int_a^b (F_y h + F_{y'} h' + \cdots + F_{y^{(n)}} h^{(n)}) \, dx = 0. \tag{20}$$

Repeatedly integrating (20) by parts and using the boundary conditions (19), we find that

$$\int_a^b \left[F_y - \frac{d}{dx} F_{y'} + \frac{d^2}{dx^2} F_{y''} - \cdots + (-1)^n \frac{d^n}{dx^n} F_{y^{(n)}} \right] h(x) \, dx = 0 \tag{21}$$

for any function h which has n continuous derivatives and satisfies (19). It follows from an obvious generalization of Lemma 1 of Sec. 3.1 that

$$F_y - \frac{d}{dx} F_{y'} + \frac{d^2}{dx^2} F_{y''} - \cdots + (-1)^n \frac{d^n}{dx^n} F_{y^{(n)}} = 0, \tag{22}$$

a result again called *Euler's equation*. Since (22) is a differential equation of order $2n$, its general solution contains $2n$ arbitrary constants, which can be determined from the boundary conditions (18).

Remark. This derivation of the Euler equation (22) is not completely rigorous, since the transition from (20) to (21) presupposes the existence of the derivatives

$$\frac{d}{dx} F_{y'}, \quad \frac{d^2}{dx^2} F_{y''}, \ldots, \quad \frac{d^n}{dx^n} F_{y^{(n)}}. \tag{23}$$

However, by a somewhat more elaborate argument, it can be shown that (20) implies (22) without this additional hypothesis. In fact, the argument in question *proves* the existence of the derivatives (23), as in Lemma 4 of Sec. 3.1.[5]

12. Variational Problems with Subsidiary Conditions

12.1. The isoperimetric problem. In the simplest variational problem considered in Chapter 1, the class of admissible curves was specified (apart from certain smoothness requirements) by conditions imposed on the end points of the curves. However, many applications of the calculus of variations lead to problems in which not only boundary conditions, but also

---

[5] Of course, this argument is unnecessary if it is known in advance that F has continuous partial derivatives up to order $n + 1$ (with respect to all its arguments).

conditions of quite a different type known as *subsidiary conditions* (synony-mously, *side conditions* or *constraints*) are imposed on the admissible curves. As an example, we first consider the *isoperimetric problem*,[6] which can be stated as follows: *Find the curve $y = y(x)$ for which the functional*

$$J[y] = \int_a^b F(x, y, y') \, dx \tag{24}$$

has an extremum, where the admissible curves satisfy the boundary conditions

$$y(a) = A, \qquad y(b) = B,$$

and are such that another functional

$$K[y] = \int_a^b G(x, y, y') \, dx \tag{25}$$

takes a fixed value l.

To solve this problem, we assume that the functions F and G defining the functionals (24) and (25) have continuous first and second derivatives in $[a, b]$ for arbitrary values of y and y'.   Then we have

THEOREM 1.[7]   *Given the functional*

$$J[y] = \int_a^b F(x, y, y') \, dx,$$

let the admissible curves satisfy the conditions

$$y(a) = A, \qquad y(b) = B, \qquad K[y] = \int_a^b G(x, y, y') \, dx = l, \tag{26}$$

where $K[y]$ is another functional, and let $J[y]$ have an extremum for $y = y(x)$.   Then, if $y = y(x)$ is not an extremal of $K[y]$, there exists a constant λ such that $y = y(x)$ is an extremal of the functional

$$\int_a^b (F + \lambda G) \, dx,$$

i.e., $y = y(x)$ satisfies the differential equation

$$F_y - \frac{d}{dx} F_{y'} + \lambda \left(G_y - \frac{d}{dx} G_{y'} \right) = 0. \tag{27}$$

Proof.   Let $J[y]$ have an extremum for the curve $y = y(x)$, subject to the conditions (26).   We choose two points x_1 and x_2 in the interval

---

[6] Originally, the isoperimetric problem referred to the following special problem (already mentioned on p. 3): *Among all closed curves of a given length l, find the curve enclosing the greatest area.*   This explains the designation "isoperimetric" = "with the same perimeter."

[7] The reader will easily recognize the analogy between this theorem and the familiar method of *Lagrange multipliers* for finding extrema of functions of several variables, subject to subsidiary conditions.   See e.g., D. V. Widder, *op. cit.*, Chap. 4, Sec. 5, especially Theorem 5.

$[a, b]$, where x_1 is arbitrary and x_2 satisfies a condition to be stated below, but is otherwise arbitrary. Then we give $y(x)$ an increment $\delta_1 y(x) + \delta_2 y(x)$, where $\delta_1 y(x)$ is nonzero only in a neighborhood of x_1, and $\delta_2 y(x)$ is nonzero only in a neighborhood of x_2. (Concerning this notation, see footnote 4, p. 41.) Using variational derivatives, we can write the corresponding increment ΔJ of the functional J in the form

$$\Delta J = \left\{ \frac{\delta F}{\delta y}\bigg|_{x=x_1} + \varepsilon_1 \right\} \Delta\sigma_1 + \left\{ \frac{\delta F}{\delta y}\bigg|_{x=x_2} + \varepsilon_2 \right\} \Delta\sigma_2, \qquad (28)$$

where

$$\Delta\sigma_1 = \int_a^b \delta_1 y(x) \, dx, \qquad \Delta\sigma_2 = \int_a^b \delta_2 y(x) \, dx$$

and $\varepsilon_1, \varepsilon_2 \to 0$ as $\Delta\sigma_1, \Delta\sigma_2 \to 0$ (see the Remark on p. 29).

We now require that the "varied" curve

$$y = y^*(x) = y(x) + \delta_1 y(x) + \delta_2 y(x)$$

satisfy the condition

$$K[y^*] = K[y].$$

Writing ΔK in a form similar to (28), we obtain

$$\Delta K = K[y^*] - K[y] = \left\{ \frac{\delta G}{\delta y}\bigg|_{x=x_1} \right. \qquad (29)$$
$$\left. + \varepsilon_1' \right\} \Delta\sigma_1 + \left\{ \frac{\delta G}{\delta y}\bigg|_{x=x_2} + \varepsilon_2' \right\} \Delta\sigma_2 = 0,$$

where $\varepsilon_1', \varepsilon_2' \to 0$ as $\Delta\sigma_1, \Delta\sigma_2 \to 0$. Next, we choose x_2 to be a point for which

$$\frac{\delta G}{\delta y}\bigg|_{x=x_2} \neq 0.$$

Such a point exists, since by hypothesis $y = y(x)$ is not an extremal of the functional K. With this choice of x_2, we can write the condition (29) in the form

$$\Delta\sigma_2 = - \left\{ \frac{\dfrac{\delta G}{\delta y}\bigg|_{x=x_1}}{\dfrac{\delta G}{\delta y}\bigg|_{x=x_2}} + \varepsilon' \right\} \Delta\sigma_1, \qquad (30)$$

where $\varepsilon' \to 0$ as $\Delta\sigma_1 \to 0$. Setting

$$\lambda = - \frac{\dfrac{\delta F}{\delta y}\bigg|_{x=x_2}}{\dfrac{\delta G}{\delta y}\bigg|_{x=x_2}},$$

and substituting (30) into the formula (28) for ΔJ, we obtain

$$\Delta J = \left\{ \frac{\delta F}{\delta y} \bigg|_{x=x_1} + \lambda \frac{\delta G}{\delta y} \bigg|_{x=x_1} \right\} \Delta \sigma_1 + \varepsilon \Delta \sigma_1, \qquad (31)$$

where $\varepsilon \to 0$ as $\Delta \sigma_1 \to 0$. This expression for ΔJ explicitly involves variational derivatives only at $x = x_1$, and the increment $h(x)$ is now just $\delta_1 y(x)$, since the "compensating increment" $\delta_2 y(x)$ has been taken into account automatically by using the condition $\Delta K = 0$. Thus, the first term in the right-hand side of (31) is the principal linear part of ΔJ, i.e., the variation of the functional J at the point x_1 is

$$\delta J = \left\{ \frac{\delta F}{\delta y} \bigg|_{x=x_1} + \lambda \frac{\delta G}{\delta y} \bigg|_{x=x_1} \right\} \Delta \sigma_1.$$

Since a necessary condition for an extremum is that $\delta J = 0$, and since $\Delta \sigma_1$ is nonzero while x_1 is arbitrary, we finally have

$$\frac{\delta F}{\delta y} + \lambda \frac{\delta G}{\delta y} = F_y - \frac{d}{dx} F_{y'} + \lambda \left(G_y - \frac{d}{dx} G_{y'} \right) = 0,$$

which is precisely equation (27). This completes the proof of the theorem.

To use Theorem 1 to solve a given isoperimetric problem, we first write the general solution of (27), which will contain two arbitrary constants in addition to the parameter λ. We then determine these three quantities from the boundary conditions $y(a) = A$, $y(b) = B$ and the subsidiary condition $K[y] = l$.

Everything just said generalizes immediately to the case of functionals depending on several functions $y_1, \ldots, y_n$ and subject to several subsidiary conditions of the form (25). In fact, suppose we are looking for an extremum of the functional

$$J[y_1, \ldots, y_n] = \int_a^b F(x, y_1, \ldots, y_n, y_1', \ldots, y_n') \, dx, \qquad (32)$$

subject to the conditions

$$y_i(a) = A_i, \quad y_i(b) = B_i \qquad (i = 1, \ldots, n) \qquad (33)$$

and

$$\int_a^b G_j(x, y_1, \ldots, y_n, y_1', \ldots, y_n') \, dx = l_j \qquad (j = 1, \ldots, k). \qquad (34)$$

In this case a necessary condition for an extremum is that

$$\frac{\partial}{\partial y_i} \left(F + \sum_{j=1}^k \lambda_j G_j \right) - \frac{d}{dx} \left\{ \frac{\partial}{\partial y_i'} \left(F + \sum_{j=1}^k \lambda_j G_j \right) \right\} = 0 \quad (i = 1, \ldots, n). \quad (35)$$

The $2n$ arbitrary constants appearing in the solution of the system (35), and the values of the k parameters $\lambda_1, \ldots, \lambda_k$, sometimes called *Lagrange multipliers*, are determined from the boundary conditions (33) and the subsidiary conditions (34). The proof of (35) is not essentially different from the proof of Theorem 1, and will not be given here.

12.2. Finite subsidiary conditions. In the isoperimetric problem, the subsidiary conditions which must be satisfied by the functions $y_1, \ldots, y_n$ are of the form (34), i.e., they are specified by functionals. We now consider a problem of a different type, which can be stated as follows: *Find the functions $y_i(x)$ for which the functional* (32) *has an extremum, where the admissible functions satisfy the boundary conditions*

$$y_i(a) = A_i, \quad y_i(b) = B_i \qquad (i = 1, \ldots, n)$$

and k "finite" subsidiary conditions $(k < n)$

$$g_j(x, y_1, \ldots, y_n) = 0 \qquad (j = 1, \ldots, k). \tag{36}$$

In other words, the functional (32) is not considered for all curves satisfying the boundary conditions (33), but only for those which lie in the $(n - k)$-dimensional manifold defined by the system (36).

For simplicity, we confine ourselves to the case $n = 2$, $k = 1$. Then we have

THEOREM 2.   *Given the functional*

$$J[y, z] = \int_a^b F(x, y, z, y', z') \, dx, \tag{37}$$

let the admissible curves lie on the surface

$$g(x, y, z) = 0 \tag{38}$$

and satisfy the boundary conditions

$$\begin{aligned} y(a) &= A_1, & y(b) &= B_1, \\ z(a) &= A_2, & z(b) &= B_2, \end{aligned} \tag{39}$$

and moreover, let $J[y]$ have an extremum for the curve

$$y = y(x), \qquad z = z(x). \tag{40}$$

Then, if g_y and g_z do not vanish simultaneously at any point of the surface (38), *there exists a function $\lambda(x)$ such that* (40) *is an extremal of the functional*

$$\int_a^b [F + \lambda(x)g] \, dx,$$

i.e., satisfies the differential equations

$$\begin{aligned} F_y + \lambda g_y - \frac{d}{dx} F_{y'} &= 0, \\ F_z + \lambda g_z - \frac{d}{dx} F_{z'} &= 0. \end{aligned} \tag{41}$$

Proof. As might be expected, the proof of this theorem closely resembles that of Theorem 1. Let $J[y, z]$ have an extremum for the curve (40), subject to the conditions (38) and (39), and let x_1 be an arbitrary point of the interval $[a, b]$. Then we give $y(x)$ an increment $\delta y(x)$ and $z(x)$ an increment $\delta z(x)$, where both $\delta y(x)$ and $\delta z(x)$ are nonzero only in a neighborhood $[\alpha, \beta]$ of x_1. Using variational derivatives, we can write the corresponding increment ΔJ of the functional $J[y, z]$ in the form

$$\Delta J = \left\{\frac{\delta F}{\delta y}\Big|_{x=x_1} + \varepsilon_1\right\}\Delta\sigma_1 + \left\{\frac{\delta F}{\delta z}\Big|_{x=x_1} + \varepsilon_2\right\}\Delta\sigma_2, \tag{42}$$

where

$$\Delta\sigma_1 = \int_a^b \delta y(x)\, dx, \qquad \Delta\sigma_2 = \int_a^b \delta z(x)\, dx,$$

and $\varepsilon_1, \varepsilon_2 \to 0$ as $\Delta\sigma_1, \Delta\sigma_2 \to 0$.

We now require that the "varied" curve

$$y = y^*(x) = y(x) + \delta y(x), \qquad z = z^*(x) = z(x) + \delta z(x)$$

satisfy the condition[8]

$$g(x, y^*, z^*) = 0.$$

In view of (38), this means that

$$\begin{aligned} 0 = \int_a^b [g(x, y^*, z^*) - g(x, y, z)]\, dx &= \int_a^b (\bar{g}_y\, \delta y + \bar{g}_z\, \delta z)\, dx \\ &= \{g_y|_{x=x_1} + \varepsilon_1'\}\Delta\sigma_1 + \{g_z|_{x=x_1} + \varepsilon_2'\}\Delta\sigma_2, \end{aligned} \tag{43}$$

where $\varepsilon_1', \varepsilon_2' \to 0$ as $\Delta\sigma_1, \Delta\sigma_2 \to 0$, and the overbar indicates that the corresponding derivatives are evaluated along certain intermediate curves. By hypothesis, either $g_y|_{x=x_1}$ or $g_z|_{x=x_1}$ is nonzero. If $g_z|_{x=x_1} \neq 0$, we can write the condition (43) in the form

$$\Delta\sigma_2 = -\left\{\frac{g_y|_{x=x_1}}{g_z|_{x=x_1}} + \varepsilon'\right\}\Delta\sigma_1, \tag{44}$$

where $\varepsilon' \to 0$ as $\Delta\sigma_1 \to 0$. Substituting (44) into the formula (42) for ΔJ, we obtain

$$\Delta J = \left\{\frac{\delta F}{\delta y}\Big|_{x=x_1} - \left(\frac{g_y}{g_z}\frac{\delta F}{\delta z}\right)\Big|_{x=x_1}\right\}\Delta\sigma_1 + \varepsilon\,\Delta\sigma_1,$$

---

[8] The existence of admissible curves $y = y^*(x)$, $z = z^*(x)$ close to the original curve $y = y(x)$, $z = z(x)$ follows from the implicit function theorem, which goes as follows: If the equation $g(x, y, z) = 0$ has a solution for $x = x_0$, $y = y_0$, $z = z_0$, if $g(x, y, z)$ and its first derivatives are continuous in a neighborhood of (x_0, y_0, z_0), and if $g_z(x_0, y_0, z_0) \neq 0$, then $g(x, y, z) = 0$ defines a unique function $z(x, y)$ which is continuous and differentiable with respect to x and y in a neighborhood of (x_0, y_0) and satisfies the condition $z(x_0, y_0) = z_0$. [There is an exactly analogous theorem for the case where $g_y(x_0, y_0, z_0) \neq 0$.] Thus, if $g_z[x, y(x), z(x)] \neq 0$ in a neighborhood of the point x_0, we can change the curve $y = y(x)$ to $y = y^*(x)$ in this neighborhood and then determine $z^*(x)$ from the relation $z^*(x) = z[x, y^*(x)]$.

where $\varepsilon \to 0$ as $\Delta\sigma_1 \to 0$. The first term in the right-hand side is the principal linear part of ΔJ, i.e., the variation of the functional J at the point x_1 is

$$\delta J = \left\{ \frac{\delta F}{\delta y}\Big|_{x=x_1} - \left(\frac{g_y}{g_z} \frac{\delta F}{\delta z}\right)\Big|_{x=x_1} \right\} \Delta\sigma_1.$$

Since a necessary condition for an extremum is that $\delta J = 0$, and since $\Delta\sigma_1$ is nonzero while x_1 is arbitrary, we finally have

$$\frac{\delta F}{\delta y} - \frac{g_y}{g_z} \frac{\delta F}{\delta z} = F_y - \frac{d}{dx} F_{y'} - \frac{g_y}{g_z}\left(F_z - \frac{d}{dx} F_{z'}\right) = 0$$

or

$$\frac{F_y - \dfrac{d}{dx} F_{y'}}{g_y} = \frac{F_z - \dfrac{d}{dx} F_{z'}}{g_z}. \tag{45}$$

Along the curve $y = y(x)$, $z = z(x)$, the common value of the ratios (45) is some function of x. If we denote this function by $-\lambda(x)$, then (45) reduces to precisely the system (41). This completes the proof of the theorem.

Remark 1. We note without proof that Theorem 2 remains valid when the class of admissible curves consists of smooth space curves satisfying the differential equation[9]

$$g(x, y, z, y', z') = 0. \tag{46}$$

More precisely, if the functional J has an extremum for a curve γ, subject to the condition (46), and if the derivatives $g_{y'}$, $g_{z'}$ do not vanish simultaneously along γ, then there exists a function $\lambda(x)$ such that γ is an integral curve of the system

$$\Phi_y - \frac{d}{dx} \Phi_{y'} = 0, \qquad \Phi_z - \frac{d}{dx} \Phi_{z'} = 0,$$

where

$$\Phi = F + \lambda G.$$

Remark 2. In a certain sense, we can consider a variational problem with a finite subsidiary condition to be a limiting case of an isoperimetric problem. In fact, if we assume that the condition (38) does not hold everywhere, but only at some fixed point

$$g(x_1, y, z) = 0,$$

we obtain a condition whose left-hand side can be regarded as a functional of y and z, i.e., a condition of the type appearing in the isoperimetric problem.

---

[9] In mechanics, conditions like (46), which contain derivatives, are called *nonholonomic constraints*, and conditions like (38) are called *holonomic constraints*.

Thus, the condition (38) can be regarded as an infinite set of conditions, each of which is a functional. As we have seen, in the isoperimetric problem the number of Lagrange multipliers $\lambda_1, \ldots, \lambda_k$ equals the number of conditions of constraint. In the same way, the function $\lambda(x)$ appearing in the problem with a finite subsidiary condition can be interpreted as a "Lagrange multiplier for each point x."

Example 1. *Among all curves of length l in the upper half-plane passing through the points* $(-a, 0)$ *and* $(a, 0)$, *find the one which together with the interval* $[-a, a]$ *encloses the largest area.* We are looking for the function $y = y(x)$ for which the integral

$$J[y] = \int_{-a}^{a} y \, dx$$

takes the largest value subject to the conditions

$$y(-a) = y(a) = 0, \qquad K[y] = \int_{-a}^{a} \sqrt{1 + y'^2} \, dx = l.$$

Thus, we are dealing with an isoperimetric problem. Using Theorem 1, we form the functional

$$J[y] + \lambda K[y] = \int_{-a}^{a} (y + \lambda \sqrt{1 + y'^2}) \, dx,$$

and write the corresponding Euler equation

$$1 + \lambda \frac{d}{dx} \frac{y'}{\sqrt{1 + y'^2}} = 0,$$

which implies

$$x + \lambda \frac{y'}{\sqrt{1 + y'^2}} = C_1. \tag{47}$$

Integrating (47), we obtain the equation

$$(x - C_1)^2 + (y - C_2)^2 = \lambda^2$$

of a family of circles. The values of C_1, C_2 and λ are then determined from the conditions

$$y(-a) = y(a) = 0, \qquad K[y] = l.$$

Example 2. *Among all curves lying on the sphere* $x^2 + y^2 + z^2 = a^2$ *and passing through two given points* (x_0, y_0, z_0) *and* (x_1, y_1, z_1), *find the one which has the least length.* The length of the curve $y = y(x)$, $z = z(x)$ is given by the integral

$$\int_{x_0}^{x_1} \sqrt{1 + y'^2 + z'^2} \, dx.$$

Using Theorem 2, we form the auxiliary functional

$$\int_{x_0}^{x_1} [\sqrt{1 + y'^2 + z'^2} + \lambda(x)(x^2 + y^2 + z^2)] \, dx,$$

and write the corresponding Euler equations

$$2y\lambda(x) - \frac{d}{dx} \frac{y'}{\sqrt{1 + y'^2 + z'^2}} = 0,$$

$$2z\lambda(x) - \frac{d}{dx} \frac{z'}{\sqrt{1 + y'^2 + z'^2}} = 0.$$

Solving these equations, we obtain a family of curves depending on four constants, whose values are determined from the boundary conditions

$$y(x_0) = y_0, \qquad y(x_1) = y_1,$$
$$z(x_0) = z_0, \qquad z(x_1) = z_1.$$

Remark. As is familiar from elementary analysis, in finding an extremum of a function of n variables subject to k constraints $(k < n)$, we can use the constraints to express k variables in terms of the other $n - k$ variables. In this way, the problem is reduced to that of finding an *unconstrained* extremum of a function of $n - k$ variables, i.e., an extremum subject to no subsidiary conditions. The situation is the same in the calculus of variations. For example, the problem of finding geodesics on a given surface can be regarded as a problem subject to a constraint, as in Example 2 of this section. On the other hand, if we express the coordinates x, y and z as functions of two parameters, we can reduce the problem to that of finding an unconstrained extremum, as in Example 2 of Sec. 9.

PROBLEMS

1. Find the extremals of the functional

$$J[y, z] = \int_0^{\pi/2} (y'^2 + z'^2 + 2yz) \, dx,$$

subject to the boundary conditions

$$y(0) = 0, \qquad y(\pi/2) = 1, \qquad z(0) = 0, \qquad z(\pi/2) = 1.$$

2. Find the extremals of the fixed end point problems corresponding to the following functionals:

a) $\int_{x_0}^{x_1} (y'^2 + z'^2 + y'z') \, dx;$

b) $\int_{x_0}^{x_1} (2yz - 2y^2 + y'^2 - z'^2) \, dx.$

3. Find the extremals of a functional of the form

$$\int_{x_0}^{x_1} F(y', z') \, dx,$$

given that $F_{y'y'}F_{z'z'} - (F_{y'z'})^2 \neq 0$ for $x_0 \leqslant x \leqslant x_1$.

Ans. A family of straight lines in three dimensions.

4. State and prove the generalization of Theorem 3 of Sec. 4.1 for functionals of the form

$$\int_a^b F(x, y_1, \ldots, y_n, y_1', \ldots, y_n')\, dx.$$

Hint. The condition $F_{y'y'} \neq 0$ is replaced by the condition $\det \|F_{y_i'y_k'}\| \neq 0$.

5. What is the condition for a functional of the form

$$\int_{t_0}^{t_1} F(t, y_1, \ldots, y_n, y_1', \ldots, y_n')\, dt,$$

depending on an n-dimensional curve $y_i = y_i(x)$, $i = 1, \ldots, n$, to be independent of the parameterization?

6. Generalizing the definition of Sec. 10, we say that the function $f(x_1, \ldots, x_n)$ is *positive-homogeneous of degree k* in $x_1, \ldots, x_n$ if

$$f(\lambda x_1, \ldots, \lambda x_n) \equiv \lambda^k f(x_1, \ldots, x_n)$$

for every $\lambda > 0$. Prove the following result, known as *Euler's theorem*: If $f(x_1, \ldots, x_n)$ is continuously differentiable and positive-homogeneous of degree k, then

$$\sum_{i=1}^n \frac{\partial f}{\partial x_i} x_i = kf.$$

7. State and prove the converse of Euler's theorem.

8. Verify formula (16) of Sec. (10).

Hint. Use Euler's theorem.

9. Prove that the Euler equations (15) of the variational problem in parametric form can be written as

$$\Phi_{x\dot{y}} - \Phi_{\dot{x}y} + (\dot{x}\ddot{y} - \ddot{x}\dot{y})\Phi_1 = 0, \tag{a}$$

where Φ_1 is a positive-homogeneous function of degree -3 satisfying the relations

$$\Phi_{\dot{x}\dot{x}} = \dot{y}^2 \Phi_1, \qquad \Phi_{\dot{x}\dot{y}} = -\dot{x}\dot{y}\Phi_1, \qquad \Phi_{\dot{y}\dot{y}} = \dot{x}^2 \Phi_1.$$

Comment. Equation (a) is known as *Weierstrass' form of the Euler equations.* It can also be written as

$$\frac{1}{\rho} = \frac{\Phi_{\dot{x}y} - \Phi_{x\dot{y}}}{\Phi_1(\dot{x}^2 + \dot{y}^2)^{3/2}},$$

where ρ is the radius of curvature of the extremal.

10. Prove that Weierstrass' form of the Euler equations is invariant under parameter changes $t = t(\tau)$, $dt/d\tau > 0$.

11. Find the extremals of the functional

$$J[y] = \int_0^1 (1 + y''^2)\, dx,$$

subject to the boundary conditions

$$y(0) = 0, \qquad y'(0) = 1, \qquad y(1) = 1, \qquad y'(1) = 1.$$

12. Find the extremals of the functional

$$J[y] = \int_0^{\pi/2} (y''^2 - y^2 + x^2)\, dx,$$

subject to the boundary conditions

$$y(0) = 1, \qquad y'(0) = 0, \qquad y(\pi/2) = 0, \qquad y'(\pi/2) = 1.$$

13. Show that the Euler equation of the functional

$$\int_{x_0}^{x_1} F(x, y, y', y'')\, dx$$

has the first integral

$$F_{y'} - \frac{d}{dx} F_{y''} = \text{const}$$

if the integrand does not depend on y, and the first integral

$$F - y'\left(F_{y'} - \frac{d}{dx} F_{y''}\right) - y''F_{y''} = \text{const}$$

if the integrand does not depend on x.

14. Find the curve joining the points $(0, 0)$ and $(1, 0)$ for which the integral

$$\int_0^1 y''^2\, dx$$

is a minimum if
   a) $y'(0) = a,\; y'(1) = b$;
   b) No other conditions are prescribed.

15. Supply the details of the argument mentioned in the remark on p. 42.

16. By direct calculation, without recourse to variational methods, prove that the isosceles triangle has the greatest area among all triangles with a given base line and a given perimeter.

Hint. All the triangles in question have the given base line and a vertex lying on a certain ellipse.

17. Find the equilibrium position of a heavy flexible inextensible cord of length l, fastened at its ends.

Hint. Minimize the ordinate of the center of gravity of the cord. By making a suitable change of variables, reduce the problem to Example 2 of Sec. 4.2.

18. Find the extremals of the functional

$$J[y] = \int_0^1 (y'^2 + x^2)\, dx,$$

subject to the conditions

$$y(0) = 0, \qquad y(1) = 0, \qquad \int_0^1 y^2\, dx = 2.$$

19. Suppose an airplane with fixed air speed v_0 makes a flight lasting T seconds. Along what closed curve should it fly if this curve is to enclose the greatest area? It is assumed that the wind velocity has constant direction and magnitude $a < v_0$.

Ans. An ellipse whose major axis is perpendicular to the wind velocity and whose eccentricity is a/v_0. The velocity of the airplane is perpendicular to the radius vector of the ellipse.

20. Given two points A and B in the xy-plane, let γ be a fixed curve joining them. Among all curves of length l joining A and B, find the curve which together with γ encloses the greatest area.

21. Generalizing the preceding problem, suppose the xy-plane is covered by a mass distribution with continuous density $\mu(x, y)$. As before, let A and B be two points in the plane, and let γ be a fixed curve joining them. Among all curves of length l joining A and B, find the curve which together with γ bounds the region of greatest mass.

Hint. Introduce the auxiliary function $V(x, y) = \int \mu(x, y) \, dx$. Then use Green's theorem and Weierstrass' form of the Euler equations.

22. Among all curves joining a given point $(0, b)$ on the y-axis to a point on the x-axis and enclosing a given area S together with the x-axis, find the curve which generates the least area when rotated about the x-axis.

Ans. The line

$$\frac{x}{a} + \frac{y}{b} = 1,$$

where $ab = 2S$.

3

THE GENERAL VARIATION
OF A FUNCTIONAL

---

13. Derivation of the Basic Formula

In this section, we derive the general formula for the variation of a functional of the form

$$J[y_1, \ldots, y_n] = \int_{x_0}^{x_1} F(x, y_1, \ldots, y_n, y_1', \ldots, y_n') \, dx, \qquad (1)$$

beginning with the case where (1) depends on a single function y and hence reduces to

$$J[y] = \int_{x_0}^{x_1} F(x, y, y') \, dx. \qquad (2)$$

We assume that all admissible curves are smooth, but, departing from our previous hypothesis, we assume that the end points of the curves for which (2) is defined can move in an arbitrary way. By the *distance* between two curves $y = y(x)$ and $y = y^*(x)$ is meant the quantity

$$\rho(y, y^*) = \max |y - y^*| + \max |y' - y^{*\prime}| + \rho(P_0, P_0^*) + \rho(P_1, P_1^*), \quad (3)$$

where P_0, P_0^* denote the left-hand end points of the curves $y = y(x)$, $y = y^*(x)$, respectively, and P_1, P_1^* denote their right-hand end points.[1] In general, the functions y and y^* are defined on different intervals I and I^*. Thus, in order for (3) to make sense, we have to extend y and y^* onto some interval containing both I and I^*. For example, this can be done by drawing tangents to the curves at their end points, as shown in Figure 4.

---

[1] In the right-hand side of (3), ρ denotes the ordinary Euclidean distance.

Now let $y = y(x)$ and $y = y^*(x)$ be two neighboring curves, in the sense of the distance (3), and let[2]

$$h(x) = y^*(x) - y(x).$$

Moreover, let

$$P_0 = (x_0, y_0), \qquad P_1 = (x_1, y_1)$$

denote the end points of the curve $y = y(x)$, while the end points of the curve $y = y^*(x) = y(x) + h(x)$ are denoted by

$$P_0^* = (x_0 + \delta x_0, y_0 + \delta y_0), \qquad P_1^* = (x_1 + \delta x_1, y_1 + \delta y_1).$$

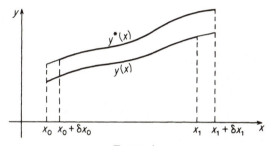

FIGURE 4

The corresponding variation δJ of the functional $J[y]$ is defined as the expression which is linear in h, h', δx_0, δy_0, δx_1, δy_1, and which differs from the increment

$$\Delta J = J[y + h] - J[y]$$

by a quantity of order higher than 1 relative to $\rho(y, y + h)$. Since[3]

$$\Delta J = \int_{x_0 + \delta x_0}^{x_1 + \delta x_1} F(x, y + h, y' + h') \, dx - \int_{x_0}^{x_1} F(x, y, y') \, dx$$

$$= \int_{x_0}^{x_1} [F(x, y + h, y' + h') - F(x, y, y')] \, dx \tag{4}$$

$$+ \int_{x_1}^{x_1 + \delta x_1} F(x, y + h, y' + h') \, dx - \int_{x_0}^{x_0 + \delta x_0} F(x, y + h, y' + h') \, dx,$$

it follows by using Taylor's theorem and letting the symbol $\sim$ denote equality except for terms of order higher than 1 relative to $\rho(y, y + h)$ that

$$\Delta J \sim \int_{x_0}^{x_1} [F_y(x, y, y')h + F_{y'}(x, y, y')h'] \, dx$$

$$+ F(x, y, y')|_{x = x_1} \delta x_1 - F(x, y, y')|_{x = x_0} \delta x_0$$

$$= \int_{x_0}^{x_1} \left[F_y - \frac{d}{dx} F_{y'} \right] h(x) \, dx + F|_{x = x_1} \delta x_1 + F_{y'} h|_{x = x_1}$$

$$- F|_{x = x_0} \delta x_0 - F_{y'} h|_{x = x_0},$$

---

[2] Note that it is no longer appropriate to write $h(x) = \delta y(x)$, as in footnote 4, p. 41. In fact, in the more precise notation of Sec. 37, $h(x) = \overline{\delta y}(x)$.

[3] Recall that we have agreed to extend $y(x)$ and $y^*(x)$ linearly onto the interval $[x_0, x_1 + \delta x_1]$, so that all integrals in the right-hand side of (4) are meaningful.

where the term containing h' has been integrated by parts. However, it is clear from Figure 4 that

$$h(x_0) \sim \delta y_0 - y'(x_0) \, \delta x_0,$$
$$h(x_1) \sim \delta y_1 - y'(x_1) \, \delta x_1,$$

where $\sim$ has the same meaning as before, and hence

$$\delta J = \int_{x_0}^{x_1} \left[F_y - \frac{d}{dx} F_{y'} \right] h(x) \, dx + F_{y'}|_{x=x_1} \, \delta y_1 + (F - F_{y'} y')|_{x=x_1} \, \delta x_1 \tag{5}$$
$$- F_{y'}|_{x=x_0} \, \delta y_0 - (F - F_{y'} y')|_{x=x_0} \, \delta x_0,$$

or more concisely,

$$\delta J = \int_{x_0}^{x_1} \left[F_y - \frac{d}{dx} F_{y'} \right] h(x) \, dx + F_{y'} \, \delta y \Big|_{x=x_0}^{x=x_1} + (F - F_{y'} y') \, \delta x \Big|_{x=x_0}^{x=x_1},$$

where we *define*

$$\delta x|_{x=x_i} = \delta x_i, \quad \delta y|_{x=x_i} = \delta y_i \qquad (i = 0, 1).$$

This is the basic formula for the general variation of the functional $J[y]$. If the end points of the admissible curves are constrained to lie on the straight lines $x = x_0$, $x = x_1$, as in the simple variable end point problem considered in Sec. 6, then $\delta x_0 = \delta x_1 = 0$, while, in the case of the fixed end point problem, $\delta x_0 = \delta x_1 = 0$ and $\delta y_0 = \delta y_1 = 0$.

Next, we return to the more general functional (1), which depends on n functions $y_1, \ldots, y_n$. Since any system of n functions can be interpreted as a curve in $(n + 1)$-dimensional Euclidean space $\mathscr{E}_{n+1}$, we can regard (1) as defined on some set of curves in $\mathscr{E}_{n+1}$. Paralleling the treatment just given for $n = 1$, we now calculate the variation of the functional (1) when there are no restrictions on the end points of the admissible curves. As before, we write

$$h_i(x) = y_i^*(x) - y_i(x) \qquad (i = 1, \ldots, n),$$

where for each i, the function $y_i^*(x)$ is close to $y_i(x)$ in the sense of the distance (3). Moreover, we let

$$P_0 = (x_0, y_1^0, \ldots, y_n^0), \qquad P_1 = (x_1, y_1^1, \ldots, y_n^1)$$

denote the end points of the curve $y_i = y_i(x)$, $i = 1, \ldots, n$, while the end points of the curve $y_i = y_i^*(x) = y_i(x) + h_i(x)$, $i = 1, \ldots, n$, are denoted by

$$P_0^* = (x_0 + \delta x_0, y_1^0 + \delta y_1^0, \ldots, y_n^0 + \delta y_n^0),$$
$$P_1^* = (x_1 + \delta x_1, y_1^1 + \delta y_1^1, \ldots, y_n^1 + \delta y_n^1),$$

and once more, we extend the functions $y_i(x)$ and $y_i^*(x)$ linearly onto the interval $[x_0, x_1 + \delta x_1]$. The corresponding variation δJ of the functional $J[y_1, \ldots, y_n]$ is defined as the expression which is linear in δx_0, δx_1 and all the quantities $h_i, h_i', \delta y_i^0, \delta y_i^1$ $(i = 1, \ldots, n)$, and which differs from the increment

$$\Delta J = J[y_1 + h_1, \ldots, y_n + h_n] - J[y_1, \ldots, y_n]$$

by a quantity of order higher than 1 relative to

$$\varrho(y_1, y_1^*) + \cdots + \varrho(y_n, y_n^*). \tag{6}$$

Since

$$\Delta J = \int_{x_0 + \delta x_0}^{x_1 + \delta x_1} F(x, \ldots, y_i + h_i, y_i' + h_i', \ldots)\, dx - \int_{x_0}^{x_1} F(x, \ldots, y_i, y_i', \ldots)\, dx$$

$$= \int_{x_0}^{x_1} [F(x, \ldots, y_i + h_i, y_i' + h_i', \ldots) - F(x, \ldots, y_i, y_i', \ldots)]\, dx$$

$$+ \int_{x_1}^{x_1 + \delta x_1} F(x, \ldots, y_i + h_i, y_i' + h_i', \ldots)$$

$$- \int_{x_0}^{x_0 + \delta x_0} F(x, \ldots, y_i + h_i, y_i' + h_i', \ldots)\, dx,$$

it follows by using Taylor's theorem and letting the symbol $\sim$ denote equality except for terms of order higher than 1 relative to the quantity (6) that

$$\Delta J \sim \int_{x_0}^{x_1} \sum_{i=1}^{n} (F_{y_i} h_i + F_{y_i'} h_i')\, dx + F|_{x=x_1}\, \delta x_1 - F|_{x=x_0}\, \delta x_0$$

$$= \int_{x_0}^{x_1} \sum_{i=1}^{n} \left(F_{y_i} - \frac{d}{dx} F_{y_i'}\right) h_i(x)\, dx + F|_{x=x_1}\, \delta x_1 + \sum_{i=1}^{n} F_{y_i'} h_i|_{x=x_1}$$

$$- F|_{x=x_0}\, \delta x_0 - \sum_{i=1}^{n} F_{y_i'} h_i|_{x=x_0},$$

where the terms containing h_i' have been integrated by parts.   Just as in the case $n = 1$, we have

$$h_i(x_0) \sim \delta y_i^0 - y_i'(x_0)\, \delta x_0,$$
$$h_i(x_1) \sim \delta y_i^1 - y_i'(x_1)\, \delta x_1,$$

and hence

$$\delta J = \int_{x_0}^{x_1} \sum_{i=1}^{n} \left(F_{y_i} - \frac{d}{dx} F_{y_i'}\right) h_i(x)\, dx$$

$$+ \sum_{i=1}^{n} F_{y_i'}\bigg|_{x=x_1} \delta y_i^1 + \left(F - \sum_{i=1}^{n} y_i' F_{y_i'}\right)\bigg|_{x=x_1} \delta x_1$$

$$- \sum_{i=1}^{n} F_{y_i'}\bigg|_{x=x_0} \delta y_i^0 - \left(F - \sum_{i=1}^{n} y_i' F_{y_i'}\right)\bigg|_{x=x_0} \delta x_0,$$

or more concisely,

$$\delta J = \int_{x_0}^{x_1} \sum_{i=1}^{n} \left(F_{y_i} - \frac{d}{dx} F_{y_i'}\right) h_i(x)\, dx$$

$$+ \sum_{i=1}^{n} F_{y_i'}\, \delta y_i \bigg|_{x=x_0}^{x=x_1} + \left(F - \sum_{i=1}^{n} y_i' F_{y_i'}\right) \delta x \bigg|_{x=x_0}^{x=x_1}, \tag{7}$$

where, as before, we *define*

$$\delta x|_{x=x_j} = \delta x_j, \quad \delta y_i|_{x=x_j} = \delta y_i^j \qquad (j = 0, 1).$$

This is the basic formula for the general variation of the functional
$J[y_1, \ldots, y_n]$.

We now write an even more concise formula for the variation (7), at
the same time introducing some important new ideas, to be discussed in
more detail in the next chapter.   Let

$$p_i = F_{y_i'} \qquad (i = 1, \ldots, n), \tag{8}$$

and suppose that the Jacobian

$$\frac{\partial(p_1, \ldots, p_n)}{\partial(y_1', \ldots, y_n')} = \det \|F_{y_i' y_k'}\|$$

is nonzero.[4]   Then we can solve the equations (8) for $y_1', \ldots, y_n'$ as functions
of the variables

$$x, y_1, \ldots, y_n, p_1, \ldots, p_n. \tag{9}$$

Next, we express the function $F(x, y_1, \ldots, y_n, y_1', \ldots, y_n')$ appearing in (1)
in terms of a new function $H(x, y_1, \ldots, y_n, p_1, \ldots, p_n)$ related to F by the
formula

$$H = -F + \sum_{i=1}^{n} y_i' F_{y_i'} \equiv -F + \sum_{i=1}^{n} y_i' p_i,$$

where the y_i' are regarded as functions of the variables (9).   The function
H is called the *Hamiltonian* (*function*) corresponding to the functional
$J[y_1, \ldots, y_n]$.   In this way, we can make a local transformation (see footnote
2, p. 68) from the "variables" $x, y_1, \ldots, y_n, y_1', \ldots, y_n', F$ appearing in (1)
to the new quantities $x, y_1, \ldots, y_n, p_1, \ldots, p_n, H$, called the *canonical
variables* (corresponding to the functional $J[y_1, \ldots, y_n]$).   In terms of the canonical variables, we can write (7) in the form

$$\delta J = \int_{x_0}^{x_1} \sum_{i=1}^{n} \left(F_{y_i} - \frac{dp_i}{dx}\right) h_i(x)\, dx + \left(\sum_{i=1}^{n} p_i\, \delta y_i - H\, \delta x\right)\Bigg|_{x=x_0}^{x=x_1}.$$

Remark.   Suppose the functional $J[y_1, \ldots, y_n]$ has an extremum (in a
certain class of admissible curves) for some curve

$$y_i = y_i(x) \qquad (i = 1, \ldots, n) \tag{10}$$

joining the points

$$P_0 = (x_0, y_1^0, \ldots, y_n^0), \qquad P_1 = (x_1, y_1^1, \ldots, y_n^1).$$

Then, since $J[y_1, \ldots, y_n]$ has an extremum for (10) compared to all admissible
curves, it certainly has an extremum for (10) compared to all curves with
fixed end points P_0 and P_1.   Therefore, (10) is an extremal, i.e., a solution
of the Euler equations

$$F_{y_i} - \frac{d}{dx} F_{y_i'} = 0 \qquad (i = 1, \ldots, n),$$

---

[4] By $\det \|a_{ik}\|$ is meant the determinant of the matrix $\|a_{ik}\|$.

so that the integral in (7) vanishes, and we are left with the formula

$$\delta J = \left[\sum_{i=1}^{n} F_{y_i'} \, \delta y_i + \left(F - \sum_{i=1}^{n} y_i' F_{y_i'} \right) \delta x \right] \Bigg|_{x = x_0}^{x = x_1}, \tag{11}$$

or in canonical variables

$$\delta J = \left(\sum_{i=1}^{n} p_i \, \delta y_i - H \, \delta x \right) \Bigg|_{x = x_0}^{x = x_1}. \tag{12}$$

Thus, regardless of the boundary conditions defining our variable end point problem, the curve for which $J[y_1, \ldots, y_n]$ has an extremum must first be an extremal and then satisfy the condition that (11) or (12) vanish (see Problem 1, p. 63).

14. End Points Lying on Two Given Curves or Surfaces

The first two chapters of this book have been devoted mainly to fixed end point problems, where the boundary conditions require that all admissible curves have two given end points. The only exception is the simple variable end point problem considered in Sec. 6, where the end points of the admissible curves are free to move along two fixed straight lines parallel to the y-axis. We now consider a more general variable end point problem. To keep matters simple, we start with the case where there is only one unknown function. Our problem can be stated as follows: *Among all smooth curves whose end points P_0 and P_1 lie on two given curves $y = \varphi(x)$ and $y = \psi(x)$, find the curve for which the functional*

$$J[y] = \int_{x_0}^{x_1} F(x, y, y') \, dx$$

has an extremum. For example, the problem of finding the distance between two plane curves is of this type, with

$$F(x, y, y') = \sqrt{1 + y'^2}.$$

As shown in the preceding section, the general variation of the functional $J[y]$ is given by formula (5). If $J[y]$ has an extremum for the curve $y = y(x)$, then, as noted at the end of Sec. 13, this curve must first of all be an extremal, i.e., a solution of Euler's equation. Hence, the integral in (5) vanishes and we have

$$\delta J = F_{y'}|_{x = x_1} \, \delta y_1 + (F - F_{y'} y')|_{x = x_1} \, \delta x_1$$
$$- F_{y'}|_{x = x_0} \, \delta y_0 - (F - F_{y'} y')|_{x = x_0} \, \delta x_0,$$

which must vanish if $J[y]$ is to have an extremum for $y = y(x)$.

Next, we observe that according to Figure 5,

$$\delta y_0 = [\varphi'(x) + \varepsilon_0]\,\delta x_0, \qquad \delta y_1 = [\psi'(x_1) + \varepsilon_1]\,\delta x_1,$$

where $\varepsilon_0 \to 0$ as $\delta x_0 \to 0$, and $\varepsilon_1 \to 0$ as $\delta x_1 \to 0$. Thus, in the present case, the condition $\delta J = 0$ becomes

$$\delta J = (F_{y'}\psi' + F - y'F_{y'})|_{x=x_1}\,\delta x_1 - (F_{y'}\varphi' + F - y'F_{y'})|_{x=x_0}\,\delta x_0 = 0,$$
(13)

since δJ contains only terms of the first order in δx_0 and δx_1. Since the increments δx_0 and δx_1 are independent, (13) implies the boundary conditions

$$(F_{y'}\varphi' + F - y'F_{y'})|_{x=x_0} = 0,$$
$$(F_{y'}\psi' + F - y'F_{y'})|_{x=x_1} = 0,$$

or

$$[F + (\varphi' - y')F_{y'}]|_{x=x_0} = 0,$$
$$[F + (\psi' - y')F_{y'}]|_{x=x_1} = 0,$$

called the *transversality conditions*. The curve $y = y(x)$ satisfying these conditions is said to be a *transversal* of the curves $y = \varphi(x)$ and $y = \psi(x)$.

Thus, to solve this kind of variable end point problem, we must first solve Euler's equation

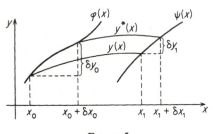

FIGURE 5

$$F_y - \frac{d}{dx}\,F_{y'} = 0, \qquad (14)$$

and then use the transversality conditions to determine the values of the two arbitrary constants appearing in the general solution of (14).

In solving variational problems, we often encounter functionals of the form

$$\int_{x_0}^{x_1} f(x, y)\sqrt{1 + y'^2}\,dx. \qquad (15)$$

For such functionals, the transversality conditions have a particularly simple appearance. In fact, in this case,

$$F_{y'} = f(x, y)\,\frac{y'}{\sqrt{1 + y'^2}} = \frac{y'F}{1 + y'^2},$$

so that the transversality conditions become

$$F + (\varphi' - y')F_{y'} = \frac{(1 + y'\varphi')F}{1 + y'^2} = 0,$$
$$F + (\psi' - y')F_{y'} = \frac{(1 + y'\psi')F}{1 + y'^2} = 0.$$

It follows that

$$y' = -\frac{1}{\varphi'},$$

at the left-hand end point, while

$$y' = -\frac{1}{\psi'}$$

at the right-hand end point, i.e., for functionals of the form (15), transversality reduces to orthogonality.

The same kind of variable end point problem can be posed for functionals depending on several functions. For example, consider the following problem: *Among all smooth curves whose end points lie on two given surfaces $x = \varphi(y, z)$ and $x = \psi(y, z)$ find the curve for which the functional*

$$J[y, z] = \int_{x_0}^{x_1} F(x, y, z, y', z') \, dx$$

has an extremum. Setting $n = 2$ in formula (7) of the preceding section, we obtain the general variation of the functional $J[y, z]$. By the same argument as in the case of one independent function, we find that the required curve $y = y(x)$, $z = z(x)$ must again be an extremal, i.e., satisfy the Euler equations

$$F_y - \frac{d}{dx} F_{y'} = 0, \qquad F_z - \frac{d}{dx} F_{z'} = 0.$$

The boundary conditions are now

$$[F_{y'} + \frac{\partial \varphi}{\partial y}(F - y'F_{y'} - z'F_{z'})]\big|_{x=x_0} = 0,$$

$$[F_{z'} + \frac{\partial \varphi}{\partial z}(F - y'F_{y'} - z'F_{z'})]\big|_{x=x_0} = 0,$$

$$[F_{y'} + \frac{\partial \psi}{\partial y}(F - y'F_{y'} - z'F_{z'})]\big|_{x=x_1} = 0,$$

$$[F_{z'} + \frac{\partial \psi}{\partial z}(F - y'F_{y'} - z'F_{z'})]\big|_{x=x_1} = 0,$$

and are again called the *transversality conditions*.

15. Broken Extremals. The Weierstrass-Erdmann Conditions

So far, we have only considered functions defined for *smooth* curves, and hence we have only permitted smooth solutions of variational problems. However, it is easy to give examples of variational problems which have no solutions in the class of smooth curves, but which have solutions if we extend the class of admissible curves to include *piecewise smooth* curves. Thus, consider the functional

$$J[y] = \int_{-1}^{1} y^2(1 - y')^2 \, dx, \qquad y(-1) = 0, \qquad y(1) = 1.$$

The greatest lower bound of the values of $J[y]$ for smooth $y = y(x)$ satisfying the boundary conditions is obviously zero, but it does not achieve this value for any smooth curve. In fact, the minimum is achieved for the curve

$$y = y(x) = \begin{cases} 0 & \text{for} \quad -1 \leqslant x \leqslant 0, \\ x & \text{for} \quad \ \ 0 < x \leqslant 1, \end{cases}$$

which has a *corner* (i.e., a discontinuous first derivative) at the point $x = 0$. Such a piecewise smooth extremal with corners is called a *broken extremal*.

Another problem involving broken extremals has already been encountered in Example 2, p. 20. There it is required to find the curve joining two points (x_0, y_0) and (x_1, y_1) which generates the surface of least area when rotated about the x-axis. As already noted, if y_0 and y_1 are sufficiently small compared to $x_1 - x_0$, the solution of the problem is given by the broken extremal Ax_0x_1B shown in Fig. 2(b), p. 21. This extremal consists of three line segments (two vertical and one horizontal) and can be included in the class of piecewise smooth curves if we set up the problem in parametric form.

Guided by the above considerations, we enlarge the class of admissible functions, relaxing the requirement that they be smooth everywhere. Thus, we pose the following problem: *Among all functions $y(x)$ which are continuously differentiable for $a \leqslant x \leqslant b$ except possibly at some point c $(a < c < b)$, and which satisfy the boundary conditions*

$$y(a) = A, \qquad y(b) = B, \tag{16}$$

find the function for which the functional

$$J[y] = \int_a^b F(x, y, y')\, dx$$

has a weak extremum. It is clear that on each of the intervals $[a, c]$ and $[c, b]$ the function for which $J[y]$ has an extremum must satisfy the Euler equation

$$F_y - \frac{d}{dx} F_{y'} = 0. \tag{17}$$

Writing $J[y]$ as a sum of two functionals, i.e.,

$$\begin{aligned}
J[y] &= \int_a^b F(x, y, y')\, dx \\
&= \int_a^c F(x, y, y')\, dx + \int_c^b F(x, y, y')\, dx \equiv J_1[y] + J_2[y],
\end{aligned}$$

we calculate the variations δJ_1 and δJ_2 of the two terms separately. The end points $x = a$, $x = b$ are fixed, and we require that the two "pieces" of the function $y(x)$ join continuously at $x = c$, but otherwise the point $x = c$ can move freely. Using formula (5) to write δJ_1 and δJ_2, and recalling that $y(x)$ is an extremal, we find that

$$\delta J_1 = F_{y'}|_{x=c-0}\, \delta y_1 + (F - y'F_{y'})|_{x=c-0}\, \delta x_1,$$
$$\delta J_2 = - F_{y'}|_{x=c+0}\, \delta y_1 - (F - y'F_{y'})|_{x=c+0}\, \delta x_1.$$

[The condition that $y(x)$ be continuous at $x = c$ implies that δJ_1 and δJ_2 involve the same increments δx_1 and δy_1.]   At an extremum we must have

$$\delta J = \delta J_1 + \delta J_2 = 0,$$

and hence

$$(F_{y'}|_{x=c-0} - F_{y'}|_{x=c+0})\, \delta y_1$$
$$+ [(F - y'F_{y'})|_{x=c-0} - (F - y'F_{y'})|_{x=c+0}]\, \delta x_1 = 0.$$

Since δx_1 and δy_1 are arbitrary, the conditions

$$F_{y'}|_{x=c-0} = F_{y'}|_{x=c+0}, \qquad (18)$$
$$(F - y'F_{y'})|_{x=c-0} = (F - y'F_{y'})|_{x=c+0},$$

called the *Weierstrass-Erdmann (corner) conditions*, hold at the point c where the extremal has a corner.

In each of the intervals $[a, c]$ and $[c, b]$, the extremal $y = y(x)$ must satisfy Euler's equation (17), i.e., a second-order differential equation. Solving these two equations, we obtain four arbitrary constants, which can then be found from the boundary conditions (16) and the Weierstrass-Erdmann conditions (18).

The Weierstrass-Erdmann conditions take a particularly simple form if we use the canonical variables

$$p = F_{y'}, \qquad H = -F + y'F_{y'}$$

introduced in Sec. 13. In fact, then the conditions (18) just mean that *the canonical variables are continuous at a point where the extremal has a corner.*

The Weierstrass-Erdmann conditions have the following simple geometric interpretation: Let x and y take fixed values, plot the value of y' along one coordinate axis, and plot the values of $F(x, y, y')$ along the other. The result is a curve, called the *indicatrix*, representing $F(x, y, y')$ as a function of y'. Then the first of the conditions (18) means that the tangents to the indicatrix at the points $y'(c - 0)$ and $y'(c + 0)$ are parallel, while the second condition, which can be written in the form

$$F|_{x=c+0} - F|_{x=c-0} = F_{y'}y'|_{x=c+0} - F_{y'}y'|_{x=c-0},$$

means that the two tangents are not only parallel, but in fact *coincide*.

PROBLEMS

1. Justify the application of Theorem 2, p. 13 to the case of variable end point problems.

2. Derive the formula for the general variation of a functional of the form

$$J[y] = \int_{x_0}^{x_1} F(x, y, y')\, dx + G(x_0, y_0, x_1, y_1).$$

3. Derive the formula for the general variation of a functional of the form

$$J[y] = \int_{x_0}^{x_1} F(x, y, y', y'')\, dx.$$

4. Find the curves for which the functional

$$J[y] = \int_0^{\pi/4} (y^2 - y'^2)\, dx,$$

can have extrema, given that $y(0) = 0$, while the right-hand end point can vary along the line $x = \pi/4$.

5. Find the curves for which the functional

$$J[y] = \int_0^{x_1} \frac{\sqrt{1 + y'^2}}{y}\, dx, \qquad y(0) = 0$$

can have extrema if
   a) The point (x_1, y_1) can vary along the line $y = x - 5$;
   b) The point (x_1, y_1) can vary along the circle $(x - 9)^2 + y^2 = 9$.

Ans. a) $y = \pm\sqrt{10x - x^2}$;   b) $y = \pm\sqrt{8x - x^2}$

6. Find the curve connecting two given circles in the (vertical) plane along which a particle falls in the shortest time under the influence of gravity.

7. Find the shortest distance between the surfaces $z = \varphi(x, y)$ and $z = \psi(x, y)$.

8. Write the transversality conditions for the functional in Prob. 2 if the end points of the admissible curves $y = y(x)$ lie on two given curves $y = \varphi(x)$ and $y = \psi(x)$.

9. Write the transversality conditions for a functional of the form

$$J[y, z] = \int_{x_0}^{x_1} f(x, y, z)\sqrt{1 + y'^2 + z'^2}\, dx$$

defined for curves whose end points lie on two given surfaces $z = \varphi(x, y)$ and $z = \psi(x, y)$. Interpret the conditions geometrically.

10. Find the curves for which the functional

$$J[y, z] = \int_0^{x_1} (y'^2 + z'^2 + 2yz)\, dx$$

can have extrema, given that $y(0) = z(0) = 0$, while the point (x_1, y_1, z_1) can vary in the plane $x = x_1$.

11. Show that for functionals of the form

$$J[y] = \int_{x_0}^{x_1} f(x, y)\sqrt{1 + y'^2}\, e^{\pm \tan^{-1} y'}\, dx,$$

the transversality conditions reduce to the requirement that the curve $y = y(x)$ intersect the curves $y = \varphi(x)$ and $y = \psi(x)$ [along which its end points vary] at an angle of $45°$.

12. Find the curves for which the functional

$$J[y] = \int_0^1 (1 + y''^2)\, dx$$

can have extrema, given that $y(0) = 0$, $y'(0) = 1$, $y(1) = 1$, while $y'(1)$ can vary arbitrarily.

13. Minimize the functional

$$J[y] = \int_{-1}^1 x^{2/3} y'^2\, dx, \qquad y(-1) = -1, \quad y(1) = 1.$$

Hint. Although the extremal $y = x^{1/3}$ has no derivative at $x = 0$, it is easily verified by direct calculation that $y = x^{1/3}$ minimizes $J[y]$.

14. Given an extremal $y = y(x)$, possibly only piecewise smooth, of the functional

$$J[y] = \int_{x_0}^{x_1} F(x, y, y')\, dx, \qquad y(x_0) = y_0, \quad y(x_1) = x_1,$$

suppose that

$$F_{y'y'}[x, y(x), z] \neq 0$$

for all finite z. Prove that $y(x)$ is then actually smooth, with a smooth derivative, in $[x_0, x_1]$.

Hint. Use Theorem 3 of Sec. 4 and the geometric interpretation of the Weierstrass-Erdmann conditions given at the end of Sec. 15.

15. Prove that the functional

$$J[y] = \int_{x_0}^{x_1} (ay'^2 + byy' + cy^2)\, dx, \qquad y(x_0) = y_0, \quad y(x_1) = y_1,$$

where $a \neq 0$, can have no broken extremals.

16. Does the functional

$$J[y] = \int_{0}^{x_1} y'^3\, dx, \qquad y(0) = 0, \quad y(x_1) = y_1$$

have broken extremals?

17. Find the extremals of the functional

$$J[y] = \int_0^4 (y' - 1)^2(y' + 1)^2\, dx, \qquad y(0) = 0, \quad y(4) = 2$$

which have just one corner.

18. Find the curve for which the functional

$$J[y] = \int_a^b F(x, y, y')\, dx, \qquad y(a) = A, \quad y(b) = B$$

has an extremum if the curve can arrive at (b, B) only after touching a given curve $y = \varphi(x)$.

19. Given a curve $y = \varphi(x)$ and two points (a, A), (b, B) lying on opposite sides of the curve, consider the functional

$$J[y] = \int_a^b F(x, y, y')\, dx, \qquad y(a) = A, \quad y(b) = B,$$

where $F(x, y, y') = F_1(x, y, y')$ on the side of the curve corresponding to (a, A), and $F(x, y, y') = F_2(x, y, y')$ on the side of the curve corresponding to (b, B). Find the curve $y = y(x)$ for which $J[y]$ has an extremum.

20. Using Fermat's principle (pp. 34, 36), specialize the results of Probs. 18 and 19 to functionals of the form

$$\int_a^b f(x, y)\sqrt{1 + y'^2}\, dx,$$

thereby deriving the familiar laws of reflection and refraction for light rays.

21. Find the curves for which the functional

$$J[y] = \int_0^{10} y'^3\, dx, \qquad y(0) = 0, \quad y(10) = 0$$

can have extrema, given that the admissible curves cannot penetrate the interior of the circle with equation

$$(x - 5)^2 + y^2 = 9.$$

$$Ans. \quad y = \begin{cases} \pm \tfrac{3}{4}x & \text{for} \quad 0 \leqslant x \leqslant \tfrac{16}{5}, \\ \pm \sqrt{9 - (x - 5)^2} & \text{for} \quad \tfrac{16}{5} \leqslant x \leqslant \tfrac{34}{5}, \\ \mp \tfrac{3}{4}(x - 10) & \text{for} \quad \tfrac{34}{5} \leqslant x \leqslant 10. \end{cases}$$

4

THE CANONICAL FORM
OF THE EULER EQUATIONS
AND RELATED TOPICS

As already remarked in Sec. 1, many physical laws can be expressed as *variational principles*, i.e., in terms of extremal properties of certain functionals. In this chapter, we shall illustrate this situation by using variational methods to study the classical mechanics of a system consisting of a finite number of particles. For example, we shall show how the trajectories in phase space of a mechanical system (which describe how the system evolves in time) can be found as the extremals of a certain functional. By using the calculus of variations, we can also find quantities connected with a given physical system which do not change as the system evolves in time. These and related ideas will be our chief concern here. First, we return to the subject of canonical variables (introduced in Sec. 13), and discuss the reduction of the Euler equations to canonical form. Appendix I (p. 208) is closely related to the subject matter of this chapter, and contains another, independent derivation of the canonical equations and the Hamilton-Jacobi equation.

16. The Canonical Form of the Euler Equations

The Euler equations corresponding to the functional

$$J[y_1, \ldots, y_n] = \int_a^b F(x, y_1, \ldots, y_n, y_1', \ldots, y_n') \, dx \qquad (1)$$

(which depends on n functions) form a system of n second-order differential equations

$$F_{y_i} - \frac{d}{dx} F_{y_i'} = 0 \qquad (i = 1, \ldots, n). \tag{2}$$

This system can be reduced (in various ways) to a system of $2n$ first-order differential equations. For example, regarding $y_1', \ldots, y_n'$ as n new functions, independent of $y_1, \ldots, y_n$, we can write (2) in the form

$$\frac{dy_i}{dx} = y_i', \quad F_{y_i} - \frac{d}{dx} F_{y_i'} = 0 \qquad (i = 1, \ldots, n), \tag{3}$$

where $y_1, \ldots, y_n, y_1', \ldots, y_n'$ are $2n$ unknown functions, and x is the independent variable.[1] However, we obtain a much more convenient and symmetric form of the Euler equations if we replace $x, y_1, \ldots, y_n, y_1', \ldots, y_n'$ by another set of variables, i.e., the *canonical variables* introduced in the preceding chapter. The reader will recall that in Sec. 13, we used the equations

$$p_i = F_{y_i'} \qquad (i = 1, \ldots, n) \tag{4}$$

to write $y_1', \ldots, y_n'$ as functions of the variables[2]

$$x, y_1, \ldots, y_n, p_1, \ldots, p_n. \tag{5}$$

Then we expressed the function $F(x, y_1, \ldots, y_n, y_1', \ldots, y_n')$ appearing in (1) in terms of a new function $H(x, y_1, \ldots, y_n, p_1, \ldots, p_n)$ related to F by the formula

$$H = -F + \sum_{i=1}^{n} y_i' p_i, \tag{6}$$

where the y_i' are regarded as functions of the variables (5). The function H is called the *Hamiltonian* (corresponding to the functional $J[y_1, \ldots, y_n]$). Finally, we introduced the new variables

$$x, y_1, \ldots, y_n, p_1, \ldots, p_n, H, \tag{7}$$

---

[1] In other words, here (and elsewhere in this chapter), we regard the y_i' as new "variables." To avoid confusion, it would be preferable to write z_i instead of y_i', but we shall adhere to the commonly accepted notation. Thus, in cases where we are concerned with the derivative of a function y_i, we shall emphasize this fact by writing dy_i/dx instead of y_i'.

[2] As already noted on p. 58, in making the transition from the variables $x, y_1, \ldots, y_n,$ $y_1', \ldots, y_n'$ to the variables $x, y_1, \ldots, y_n, p_1, \ldots, p_n$, we require that the Jacobian

$$\frac{\partial(p_1, \ldots, p_n)}{\partial(y_1', \ldots, y_n')} = \det \|F_{y_i' y_k'}\|$$

be nonzero. We shall assume that this condition is satisfied. However, it should be kept in mind that this condition guarantees only the *local* "solvability" of the equations (4) with respect to $y_1', \ldots, y_n'$, but it does not guarantee the possibility of representing $y_1', \ldots, y_n'$ as functions of $x, y_1, \ldots, y_n, p_1, \ldots, p_n$ which are defined over the whole region under discussion. Thus, all our considerations have a *local* character.

called the canonical variables (corresponding to the functional $J[y_1, \ldots, y_n]$), which were used on p. 58 to write a concise expression for the general variation of the functional $J[y_1, \ldots, y_n]$, and on p. 63 to give a simple interpretation of the Weierstrass-Erdmann conditions.

We now show how the Euler equations (3) transform when we go over to canonical variables. In order to make this change in the Euler equations, we have to express the partial derivatives F_{y_i} (i.e., the partial derivatives of F with respect to y_i, evaluated for constant $x, y_1', \ldots, y_n'$) in terms of the partial derivatives H_{y_i} (evaluated for constant $x, p_1, \ldots, p_n$).[3] The direct evaluation of these derivatives would be rather formidable. Therefore, to avoid lengthy calculations, we write the expression for the differential of the function H. Then, using the fact that the first differential of a function does not depend on the choice of independent variables (i.e., is invariant under changes of the independent variables), we shall obtain the required formulas quite easily.

By the definition of H, we have

$$dH = - dF + \sum_{i=1}^{n} p_i \, dy_i' + \sum_{i=1}^{n} y_i' \, dp_i,$$

so that

$$dH = - \frac{\partial F}{\partial x} \, dx - \sum_{i=1}^{n} \frac{\partial F}{\partial y_i} \, dy_i - \sum_{i=1}^{n} \frac{\partial F}{\partial y_i'} \, dy_i' \\ + \sum_{i=1}^{n} p_i \, dy_i' + \sum_{i=1}^{n} y_i' \, dp_i. \tag{8}$$

Ordinarily, before using (8) to obtain expressions for the partial derivatives of H, we would have to express the dy_i' in terms of x, y_i, and p_i. However (and this is the important feature of the canonical variables), because of the relations

$$\frac{\partial F}{\partial y_i'} = p_i \qquad (i = 1, \ldots, n),$$

the terms containing dy_i' in (8) cancel each other out, and we obtain

$$dH = - \frac{\partial F}{\partial x} \, dx - \sum_{i=1}^{n} \frac{\partial F}{\partial y_i} \, dy_i + \sum_{i=1}^{n} y_i' \, dp_i. \tag{9}$$

Thus, to obtain the partial derivatives of H, we need only write down the appropriate coefficients of the differentials in the right-hand side of (9), i.e.,

$$\frac{\partial H}{\partial x} = - \frac{\partial F}{\partial x}, \qquad \frac{\partial H}{\partial y_i} = - \frac{\partial F}{\partial y_i}, \qquad \frac{\partial H}{\partial p_i} = y_i'.$$

---

[3] The notation ordinarily used in analysis to denote partial derivatives suffers from the familiar defect of not specifying just which variables are held fixed.

In other words, the quantities $\partial F/\partial y_i$ and y_i' are connected with the partial derivatives of the function H by the formulas

$$y_i' = \frac{\partial H}{\partial p_i}, \qquad \frac{\partial F}{\partial y_i} = -\frac{\partial H}{\partial y_i}. \tag{10}$$

Finally, using (10), we can write the Euler equations (3) in the form

$$\frac{dy_i}{dx} = \frac{\partial H}{\partial p_i}, \qquad \frac{dp_i}{dx} = -\frac{\partial H}{\partial y_i} \qquad (i = 1, \dots, n). \tag{11}$$

These $2n$ first-order differential equations form a system which is equivalent to the system (3) and is called the *canonical system of Euler equations* (or simply the *canonical Euler equations*) for the functional (1).

17. First Integrals of the Euler Equations

It will be recalled that a *first integral* of a system of differential equations is a function which has a constant value along each integral curve of the system. We now look for first integrals of the canonical system (11), and hence of the original system (3) which is equivalent to (11).   First, we consider the case where the function F defining the functional (1) does not depend on x explicitly, i.e., is of the form $F(y_1, \dots, y_n, y_1', \dots, y_n')$.   Then the function

$$H = -F + \sum_{i=1}^{n} y_i' p_i$$

also does not depend on x explicitly, and hence

$$\frac{dH}{dx} = \sum_{i=1}^{n} \left(\frac{\partial H}{\partial y_i} \frac{dy_i}{dx} + \frac{\partial H}{\partial p_i} \frac{dp_i}{dx} \right). \tag{12}$$

Using the Euler equations in the canonical form (11), we find that (12) becomes

$$\frac{dH}{dx} = \sum_{i=1}^{n} \left(\frac{\partial H}{\partial y_i} \frac{\partial H}{\partial p_i} - \frac{\partial H}{\partial p_i} \frac{\partial H}{\partial y_i} \right) = 0,$$

along each extremal.[4]   *Thus, if F does not depend on x explicitly, the function* $H(y_1, \dots, y_n, p_1, \dots, p_n)$ *is a first integral of the Euler equations.*[5]

_____

[4] If H depends on x explicitly, the formula

$$\frac{dH}{dx} = \frac{\partial H}{\partial x}$$

can be derived by the same argument.

[5] Cf. the discussion in Case 2, p. 18 of the integration of Euler's equation for functionals which are independent of x.

Next, we consider an arbitrary function of the form

$$\Phi = \Phi(y_1, \ldots, y_n, p_1, \ldots, p_n),$$

and we examine the conditions under which Φ will be a first integral of the system (11).   We drop the assumption that F does not depend on x explicitly, and instead we consider the general case.   Along each integral curve of the system (11), we have

$$\frac{d\Phi}{dx} = \sum_{i=1}^{n} \frac{\partial \Phi}{\partial y_i} \frac{dy_i}{dx} + \frac{\partial \Phi}{\partial p_i} \frac{dp_i}{dx}$$

$$= \sum_{i=1}^{n} \frac{\partial \Phi}{\partial y_i} \frac{\partial H}{\partial p_i} - \frac{\partial \Phi}{\partial p_i} \frac{\partial H}{\partial y_i} = [\Phi, H],$$

where the expression

$$[\Phi, H] = \sum_{i=1}^{n} \frac{\partial \Phi}{\partial y_i} \frac{\partial H}{\partial p_i} - \frac{\partial \Phi}{\partial p_i} \frac{\partial H}{\partial y_i}$$

is called the *Poisson bracket* of the functions Φ and H.   Thus, we have proved the formula

$$\frac{d\Phi}{dx} = [\Phi, H]. \tag{13}$$

It follows from (13) that *a necessary and sufficient condition for a function* $\Phi = \Phi(y_1, \ldots, y_n, p_1, \ldots, p_n)$ *to be a first integral of the system of Euler equations* (11) *is that the Poisson bracket* $[\Phi, H]$ *vanish identically.*[6]

18. The Legendre Transformation

We now consider another method of reducing the Euler equations to canonical form, a method which differs from that presented in Sec. 16. The idea of this new method is to replace the variational problem under consideration by another, equivalent problem, such that the Euler equations for the new problem are the same as the *canonical* Euler equations for the original problem.

18.1. We begin by discussing some related topics from the theory of extrema of functions of n variables.   First, we consider the case $n = 1$.

---

[6] According to the existence theorem for the system (11), there is an integral curve of the system passing through any given point $(x, y_1, \ldots, y_n, p_1, \ldots, p_n)$.   Hence, if $[\Phi, H] = 0$ along every integral curve, it follows that $[\Phi, H] \equiv 0$.   If Φ (as well as H) depends on x explicitly, it is easily verified that (13) is replaced by

$$\frac{d\Phi}{dx} = \frac{\partial \Phi}{\partial x} + [\Phi, H].$$

Suppose we are looking for an extremum, say a minimum, of the function $f(\xi)$, and suppose $f(\xi)$ is (*strictly*) *convex*, which means that

$$f''(\xi) > 0$$

wherever $f(\xi)$ is defined.   We introduce a new independent variable

$$p = f'(\xi), \tag{14}$$

called the *tangential coordinate*, which is just the slope of the tangent passing through a given point of the curve $\eta = f(\xi)$.   Since by hypothesis

$$\frac{dp}{d\xi} = f''(\xi) > 0,$$

we can use (14) to express ξ in terms of p.   In fact, since the function $f(\xi)$ is convex, any point of the curve $\eta = f(\xi)$ is uniquely determined by the slope of its tangent (see Figure 6).   Of course, the same is true for a (*strictly*) *concave* function, i.e., a function such that $f''(\xi) < 0$ everywhere.

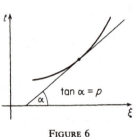

We now introduce the new function

$$H(p) = -f(\xi) + p\xi, \tag{15}$$

where ξ is regarded as the function of p obtained by solving (14).   The transformation from the variable and function pair ξ, $f(\xi)$ to the variable and function pair p, $H(p)$, defined by formulas

FIGURE 6

(14) and (15), is called the *Legendre transformation*.   It is easy to see that since $f(\xi)$ is convex, so is $H(p)$.   [The convex functions $H(p)$ and $f(\xi)$ are sometimes said to be conjugate.]   In fact,

$$dH = -f'(\xi)\,d\xi + p\,d\xi + \xi\,dp$$

implies that

$$\frac{dH}{dp} = \xi, \tag{16}$$

and hence

$$\frac{d^2H}{dp^2} = \frac{d\xi}{dp} = \frac{1}{\dfrac{dp}{d\xi}} = \frac{1}{f''(\xi)} > 0, \tag{17}$$

since $f''(\xi) > 0$.   Moreover, if the Legendre transformation is applied to the pair p, $H(p)$, we get back the pair ξ, $f(\xi)$.   This follows from (16) and the relation

$$-H(p) + pH'(p) = f(\xi) - pH'(p) + pH'(p) = f(\xi). \tag{18}$$

Thus, the Legendre transformation is an *involution*, i.e., a transformation which is its own inverse.

Example. If

$$f(\xi) = \frac{\xi^a}{a} \qquad (a > 1),$$

then

$$f'(\xi) = p = \xi^{a-1},$$

i.e.,

$$\xi = p^{1/(a-1)}.$$

It follows that

$$H = -\frac{\xi^a}{a} + p\xi = -\frac{p^{a/(a-1)}}{a} + pp^{1/(a-1)/(a-1)} = p^{a/(a-1)}\left(-\frac{1}{a} + 1\right),$$

and therefore

$$H(p) = \frac{p^b}{b},$$

where b is related to a by the formula

$$\frac{1}{a} + \frac{1}{b} = 1.$$

Next, we show that if

$$-H(p) + \xi p \tag{19}$$

is regarded as a function of two variables, then

$$f(\xi) = \max_p \left[-H(p) + \xi p\right]. \tag{20}$$

[In fact, we can use (20) instead of (15) to define the function $H(p)$.] To prove this result, we note that according to (18), the function (19) reduces to $f(\xi)$ when the condition

$$\frac{\partial}{\partial p}\left[-H(p) + \xi p\right] = -H'(p) + \xi = 0,$$

or

$$\xi = H'(p),$$

is satisfied. Thus, $f(\xi)$ is an extremum of the function $-H(p) + \xi p$, regarded as a function of p. Moreover, the extremum is a maximum since

$$\frac{\partial^2}{\partial p^2}\left[-H(p) + \xi p\right] = -H''(p) < 0$$

[cf. (17)]. It follows that

$$\min_\xi f(\xi) = \min_\xi \max_p \left[-H(p) + \xi p\right],$$

i.e., the extremum of $f(\xi)$ is also an extremum of (19), regarded as a function of two variables.

Similar considerations apply to functions of several independent variables. Let

$$f(\xi_1, \ldots, \xi_n)$$

be a function of n variables such that

$$\det \|f_{\xi_i \xi_k}\| \neq 0, \tag{21}$$

and let

$$p_i = f_{\xi_i} \qquad (i = 1, \ldots, n). \tag{22}$$

Then, using (22) to write $\xi_1, \ldots, \xi_n$ in terms of $p_1, \ldots, p_n$, we form the function

$$H(p_1, \ldots, p_n) = -f + \sum_{i=1}^{n} \xi_i p_i.$$

As in the case of one variable, it can be shown that

$$f(\xi_1, \ldots, \xi_n) = \operatorname*{ext}_{p_1, \ldots, p_n} \left[-H(p_1, \ldots, p_n) + \sum_{i=1}^{n} p_i \xi_i \right]$$

and

$$\operatorname*{ext}_{\xi_1, \ldots, \xi_n} f(\xi_1, \ldots, \xi_n) = \operatorname*{ext}_{\xi_1, \ldots, \xi_n, p_1, \ldots, p_n} \left[-H(p_1, \ldots, p_n) + \sum_{i=1}^{n} p_i \xi_i \right],$$

where ext denotes the operation of taking an extremum with respect to the indicated variables. In other words, the extremum of $f(\xi_1, \ldots, \xi_n)$ is also an extremum of

$$-H(p_1, \ldots, p_n) + \sum_{i=1}^{n} p_i \xi_i,$$

regarded as a function of $2n$ variables.

Remark. If instead of (21), we impose the stronger condition that the matrix

$$\|f_{\xi_i \xi_k}\|$$

be *positive definite*, i.e., that the quadratic form

$$\sum_{i, k=1}^{n} f_{\xi_i \xi_k} \alpha_i \alpha_k$$

be positive for arbitrary real numbers $\alpha_1, \ldots, \alpha_n$,[7] then

$$f(\xi_1, \ldots, \xi_n) = \operatorname*{max}_{p_1, \ldots, p_n} \left[-H(p_1, \ldots, p_n) + \sum_{i=1}^{n} \xi_i p_i \right]. \tag{23}$$

---

[7] This is the condition for the function $f(\xi_1, \ldots, \xi_n)$ to be (strictly) convex.

It follows from (23) that

$$-H(p_1, \ldots, p_n) + \sum_{i=1}^{n} p_i \xi_i \leqslant f(\xi_1, \ldots, \xi_n)$$

for arbitrary $p_1, \ldots, p_n$, i.e.,

$$\sum_{i=1}^{n} p_i \xi_i \leqslant H(p_1, \ldots, p_n) + f(\xi_1, \ldots, \xi_n),$$

a result known as *Young's inequality*.

18.2. We now apply the considerations of Sec. 18.1 to functionals. Given a functional

$$J[y] = \int_a^b F(x, y, y') \, dx, \tag{24}$$

we set

$$p = F_{y'}(x, y, y') \tag{25}$$

and

$$H(x, y, p) = -F + py'. \tag{26}$$

Here we assume that $F_{y'y'} \neq 0$, so that (25) defines y' as a function of x, y and p. Then we introduce the new functional

$$J[y, p] = \int_a^b [-H(x, y, p) + py'] \, dx, \tag{27}$$

where y and p are regarded as two independent functions, and y' is the derivative of y. This functional is obviously the same as the original functional (24), if we choose p to be given by the expression (25). The Euler equations for the functional (27) are

$$-\frac{\partial H}{\partial y} - \frac{dp}{dx} = 0, \qquad -\frac{\partial H}{\partial p} + \frac{dy}{dx} = 0, \tag{28}$$

i.e., just the canonical equations for the functional (24). If we can show that the functionals (24) and (27) have their extrema for the same curves, this will prove that the equation

$$\frac{\partial F}{\partial y} - \frac{d}{dx} \frac{\partial F}{\partial y'} = 0 \tag{29}$$

and the equations (28) are equivalent, thereby providing a new derivation of the canonical equations, independent of the derivation given in Sec. 16.

First, we observe that the transformation from the variables x, y, y' and the function F to the variables x, y, p and the function H, defined by formulas

(25) and (26), is an involution, i.e., if we subject $H(x, y, p)$ to a Legendre transformation, we get back the function $F(x, y, y')$. In fact, since

$$dH = -\frac{\partial F}{\partial x} dx - \frac{\partial F}{\partial y} dy + y' \, dp,$$

it follows that

$$\frac{\partial H}{\partial p} = y',$$

and hence

$$-H + p \frac{\partial H}{\partial p} = F - py' + py' = F. \tag{30}$$

[Cf. formula (9) of Sec. 16.]

Next, we note that to prove the equivalence of the variational problems (24) and (27), it is sufficient to show that $J[y]$ is an extremum of $J[y, p]$ when p is varied and y is held fixed, symbolically

$$J[y] = \underset{p}{\text{ext}} \, J[y, p], \tag{31}$$

since then an extremum of $J[y, p]$ when both p and y are varied will be an extremum of $J[y]$. Since $J[y, p]$ does not contain p', to find an extremum of $J[y, p]$ it is sufficient to find an extremum of the integrand in (27) at every point (cf. Case 3, p. 19). Thus we have

$$\frac{\partial}{\partial p} \left[-H + py'\right] = 0,$$

from which it follows that

$$y' = \frac{\partial H}{\partial p}.$$

But this implies (31), since

$$-H + p \frac{\partial H}{\partial p} = F,$$

according to (30). Thus, we have proved the equivalence of the variational problems (24) and (27), and of the corresponding Euler equations (28) and (29). Although we have only considered functionals depending on a single function, completely analogous considerations apply to the case of functionals depending on several functions.

Example. Consider the functional

$$\int_a^b (Py'^2 + Qy^2) \, dx, \tag{32}$$

where P and Q are functions of x. In this case,

$$p = 2Py', \qquad H = Py'^2 - Qy^2,$$

and hence

$$H = \frac{p^2}{4P} - Qy^2.$$

The corresponding canonical equations are

$$\frac{dp}{dx} = 2Qy, \qquad \frac{dy}{dx} = \frac{p}{2P},$$

while the usual form of the Euler equation for the functional (32) is

$$2yQ - \frac{d}{dx}(2Py') = 0.$$

19. Canonical Transformations

Next, we look for transformations under which the canonical Euler equations preserve their canonical form. The reader will recall that in Sec. 8 we proved the invariance of the Euler equation

$$F_y - \frac{d}{dx} F_{y'} = 0$$

under coordinate transformations of the form

$$u = u(x, y), \qquad \begin{vmatrix} u_x & u_y \\ v_x & v_y \end{vmatrix} \neq 0.$$
$$v = v(x, y),$$

(Such transformations change y' to dv/du in the original functional.) The canonical Euler equations also have this invariance property. Furthermore, because of the symmetry between the variables y_i and p_i in the canonical equations, they permit even more general changes of variables, i.e., we can transform the variables x, y_i, p_i into new variables x,

$$Y_i = Y_i(x, y_1, \ldots, y_n, p_1, \ldots, p_n),$$
$$P_i = P_i(x, y_1, \ldots, y_n, p_1, \ldots, p_n). \qquad (33)$$

In other words, we can think of letting the p_i transform according to their own formulas, independently of how the variables y_i transform. However, the canonical equations do not preserve their form under all transformations (33). We now study the conditions which have to be imposed on the transformations (33) if the Euler equations are to continue to be in canonical form when written in the new variables, i.e., if the canonical equations are to transform into new equations

$$\frac{dY_i}{dx} = \frac{\partial H^*}{\partial P_i}, \qquad \frac{dP_i}{dx} = -\frac{\partial H^*}{\partial Y_i}, \qquad (34)$$

where $H^* = H^*(x, Y_1, \ldots, Y_n, P_1, \ldots, P_n)$ is some new function. Transformations of the form (33) which preserve the canonical form of the Euler equations are called *canonical transformations*.

To find such canonical transformations, we use the fact that the canonical equations

$$\frac{dy_i}{dx} = \frac{\partial H}{\partial p_i}, \qquad \frac{dp_i}{dx} = -\frac{\partial H}{\partial y_i} \qquad (35)$$

are the Euler equations of the functional

$$J[y_1, \ldots, y_n, p_1, \ldots, p_n] = \int_a^b \left(\sum_{i=1}^n p_i y_i' - H \right) dx, \qquad (36)$$

in which the y_i and p_i are regarded as $2n$ independent functions. We want the new variables Y_i and P_i to satisfy the equations (34) for some function H^*. This suggests that we write the functional which has (34) as its Euler equations. This functional is

$$J^*[Y_1, \ldots, Y_n, P_1, \ldots, P_n] = \int_a^b \left(\sum_{i=1}^n P_i Y_i' - H^* \right) dx, \qquad (37)$$

where Y_i and P_i are the functions of x, y_i and p_i defined by (33), and Y_i' is the derivative of Y_i. Thus, the functionals (36) and (37) represent two different variational problems involving the same variables y_i and p_i, and the requirement that the new system of canonical equations (34) be equivalent to the old system (35), i.e., that it be possible to obtain (34) from (35) by making a change of variables (33), is the same as the requirement that the variational problems corresponding to the functionals (36) and (37) be equivalent.

In the remarks made on p. 36, it was shown that two variational problems are equivalent (i.e., have the same extremals) if the integrands of the corresponding functionals differ from each other by a total differential, which in this case means that

$$\sum_{i=1}^n p_i \, dy_i - H \, dx = \sum_{i=1}^n P_i \, dY_i - H^* \, dx + d\Phi(x, y_1, \ldots, y_n, p_1, \ldots, p_n)$$
$$(38)$$

for some function Φ. Thus, if a given transformation (33) from the variables x, y_i, p_i to the variables x, Y_i, P_i is such that there exists a function Φ satisfying the condition (38), then the transformation (33) is canonical. In this case, the function Φ defined by (38) is called the *generating function* of the canonical transformation. The function Φ is only specified to within an additive constant, since, as is well known, a function is only specified by its total differential to within an additive constant.

To justify the term "generating function," we must show how to actually find the canonical transformation corresponding to a given generating function Φ. This is easily done. Writing (38) in the form

$$d\Phi = \sum_{i=1}^n p_i \, dy_i - \sum_{i=1}^n P_i \, dY_i + (H^* - H) \, dx,$$

we find that[8]

$$p_i = \frac{\partial \Phi}{\partial y_i}, \qquad P_i = \frac{\partial \Phi}{\partial Y_i}, \qquad H^* = H + \frac{\partial \Phi}{\partial x}. \qquad (39)$$

---

[8] Φ is originally a function of x, y_i and p_i. However, by using (33), we can write Φ as a function of the variables x, y_i and Y_i.

Then (39) is precisely the desired canonical transformation. In fact, the $2n + 1$ equations (39) establish the connection between the old variables y_i, p_i and the new variables Y_i, P_i, and they also give an expression for the new Hamiltonian H^*. Moreover, it is obvious that (39) satisfies the condition (38), so that the transformation (38) is indeed canonical. If the generating function Φ does not depend on x explicitly, then $H^* = H$. In this case, to obtain the new Hamiltonian H^*, we need only replace y_i and p_i in H by their expressions in terms of Y_i and P_i.[9]

In writing (39), we assumed that the generating function is specified as a function of x, the old variables y_i and the new variables Y_i:

$$\Phi = \Phi(x, y_1, \ldots, y_n, Y_1, \ldots, Y_n).$$

It may be more convenient to express the generating function in terms of y_i and P_i instead of y_i and Y_i. To this end, we rewrite (38) in the form

$$d\left(\Phi + \sum_{i=1}^n P_i Y_i\right) = \sum_{i=1}^n p_i \, dy_i + \sum_{i=1}^n Y_i \, dP_i + (H^* - H) \, dx,$$

thereby obtaining a new generating function

$$\Phi + \sum_{i=1}^n P_i Y_i, \tag{40}$$

which is to be regarded as a function of the variables x, y_i and P_i. Denoting (40) by $\Psi(x, y_1, \ldots, y_n, P_1, \ldots, P_n)$, we can write the corresponding canonical transformation in the form

$$p_i = \frac{\partial \Psi}{\partial y_i}, \qquad Y_i = \frac{\partial \Psi}{\partial P_i}, \qquad H^* = H + \frac{\partial \Psi}{\partial x}. \tag{41}$$

20. Noether's Theorem

In Sec. 17 we proved that the system of Euler equations corresponding to the functional

$$\int_a^b F(y_1, \ldots, y_n, y_1', \ldots, y_n') \, dx, \tag{42}$$

where F does not depend on x explicitly, has the first integral

$$H = -F + \sum_{i=1}^n y_i' F_{y_i'}.$$

It is clear that the statement "F does not depend on x explicitly" is equivalent to the statement "F, and hence the integral (42), remains the same if we replace x by the new variable

$$x^* = x + \varepsilon, \tag{43}$$

---

[9] A similar remark holds for the function Ψ in (41).

where ε is an arbitrary constant." It follows that H is a first integral of the system of Euler equations corresponding to the functional (42) if and only if (42) is invariant under the transformation (43).[10]

We now show that even in the general case, there is a connection between the existence of certain first integrals of a system of Euler equations and the invariance of the corresponding functional under certain transformations of the variables $x, y_1, \ldots, y_n$. We begin by defining more precisely what is meant by the invariance of a functional under some set of transformations. Suppose we are given a functional[11]

$$J[y_1, \ldots, y_n] = \int_{x_0}^{x_1} F(x, y_1, \ldots, y_n, y_1', \ldots, y_n') \, dx,$$

which we write in the concise form

$$J[y] = \int_{x_0}^{x_1} F(x, y, y') \, dx, \tag{44}$$

where now y indicates the n-dimensional vector $(y_1, \ldots, y_n)$ and y' the n-dimensional vector $(y_1', \ldots, y_n')$. Consider the transformation

$$\begin{aligned}
x^* &= \Phi(x, y_1, \ldots, y_n, y_1', \ldots, y_n') = \Phi(x, y, y'), \\
y_i^* &= \Psi_i(x, y_1, \ldots, y_n, y_1', \ldots, y_n') = \Psi_i(x, y, y'),
\end{aligned} \tag{45}$$

where $i = 1, \ldots, n$. The transformation (45) carries the curve γ, with the vector equation

$$y = y(x) \qquad (x_0 \leqslant x \leqslant x_1),$$

into another curve γ^*. In fact, replacing y, y' in (45) by $y(x), y'(x)$, and eliminating x from the resulting $n + 1$ equations, we obtain the vector equation

$$y^* = y^*(x^*) \qquad (x_0^* \leqslant x^* \leqslant x_1^*)$$

for γ^*, where $y^* = (y_1^*, \ldots, y_n^*)$.

DEFINITION. *The functional* (44) *is said to be invariant under the transformation* (45) *if* $J[\gamma^*] = J[\gamma]$, *i.e., if*

$$\int_{x_0^*}^{x_1^*} F\left(x^*, y^*, \frac{dy^*}{dx^*}\right) dx^* = \int_{x_0}^{x_1} F\left(x, y, \frac{dy}{dx}\right) dx.$$

---

[10] The fact that H is a first integral *only if* (42) is invariant under the transformation (43) follows from the formula

$$\frac{dH}{dx} = \frac{\partial H}{\partial x}$$

(see footnote 4, p. 70), since $\partial H/\partial x = 0$ only if $\partial F/\partial x = 0$.

[11] To avoid confusion in what follows, the reader should note that the subscripts can play two different roles; when indexing x, they refer to different *values*, while when indexing y, they refer to different *functions*. For example, the y_i^* are new functions, while x_0^* and x_1^* are the new positions of the end points of the interval $[x_0, x_1]$.

Example 1. The functional

$$J[y] = \int_{x_0}^{x_1} y'^2 \, dx$$

is invariant under the transformation

$$x^* = x + \varepsilon, \qquad y^* = y, \tag{46}$$

where ε is an arbitrary constant. In fact, given a curve γ with equation

$$y = y(x) \qquad (x_0 \leqslant x \leqslant x_1),$$

the "transformed" curve γ^*, i.e., the curve obtained from γ by shifting it a distance ε along the x-axis, has the equation

$$y^* = y(x^* - \varepsilon) = y^*(x^*) \qquad (x_0 + \varepsilon \leqslant x^* \leqslant x_1 + \varepsilon),$$

and then

$$J[\gamma^*] = \int_{x_0^*}^{x_1^*} \left[\frac{dy^*(x^*)}{dx^*} \right]^2 dx^* = \int_{x_0+\varepsilon}^{x_1+\varepsilon} \left[\frac{dy(x^* - \varepsilon)}{dx^*} \right]^2 dx^*$$
$$= \int_{x_0}^{x_1} \left[\frac{dy(x)}{dx} \right]^2 dx = J[\gamma].$$

Example 2. The integral

$$J[y] = \int_{x_0}^{x_1} xy'^2 \, dx$$

is an example of a functional which is not invariant under the transformation (46). In fact, carrying out the same calculations as in Example 1, we obtain

$$J[\gamma^*] = \int_{x_0^*}^{x_1^*} x^* \left[\frac{dy^*(x^*)}{dx^*} \right]^2 dx^* = \int_{x_0+\varepsilon}^{x_1+\varepsilon} x^* \left[\frac{dy(x^* - \varepsilon)}{dx^*} \right]^2 dx^*$$
$$= \int_{x_0}^{x_1} (x + \varepsilon) \left[\frac{dy(x)}{dx} \right]^2 dx = J[\gamma] + \varepsilon \int_{x_0}^{x_1} \left[\frac{dy(x)}{dx} \right]^2 dx \neq J[\gamma].$$

Suppose now that we have a family of transformations

$$\begin{aligned} x^* &= \Phi(x, y, y'; \varepsilon), \\ y_i^* &= \Psi_i(x, y, y'; \varepsilon), \end{aligned} \tag{47}$$

depending on a parameter ε, where the functions Φ and Ψ_i ($i = 1, \ldots, n$) are differentiable with respect to ε, and the value $\varepsilon = 0$ corresponds to the identity transformation:

$$\begin{aligned} \Phi(x, y, y'; 0) &= x, \\ \Psi_i(x, y, y'; 0) &= y_i. \end{aligned} \tag{48}$$

Then we have the following result:

THEOREM (*Noether*). *If the functional*

$$J[y] = \int_{x_0}^{x_1} F(x, y, y') \, dx \tag{49}$$

is invariant under the family of transformations (47) *for arbitrary* x_0 *and* x_1, *then*

$$\sum_{i=1}^{n} F_{y_i'} \psi_i + \left(F - \sum_{i=1}^{n} y_i' F_{y_i'} \right) \varphi = \text{const} \tag{50}$$

along each extremal of $J[y]$, *where*

$$\varphi(x, y, y') = \frac{\partial \Phi(x, y, y'; \varepsilon)}{\partial \varepsilon} \bigg|_{\varepsilon=0},$$
$$\psi_i(x, y, y') = \frac{\partial \Psi_i(x, y, y'; \varepsilon)}{\partial \varepsilon} \bigg|_{\varepsilon=0}. \tag{51}$$

In other words, every one-parameter family of transformations leaving $J[y]$ *invariant leads to a first integral of its system of Euler equations.*

Proof. Suppose ε is a small quantity. Then, by Taylor's theorem, we have[12]

$$x^* = \Phi(x, y, y'; \varepsilon) = \Phi(x, y, y'; 0) + \varepsilon \frac{\partial \Phi(x, y, y'; \varepsilon)}{\partial \varepsilon} \bigg|_{\varepsilon=0} + o(\varepsilon),$$

$$y_i^* = \Psi_i(x, y, y'; \varepsilon) = \Psi_i(x, y, y'; 0) + \varepsilon \frac{\partial \Psi_i(x, y, y'; \varepsilon)}{\partial \varepsilon} \bigg|_{\varepsilon=0} + o(\varepsilon),$$

or using (48) and (51),

$$x^* = x + \varepsilon\varphi(x, y, y') + o(\varepsilon),$$
$$y_i^* = y_i + \varepsilon\psi_i(x, y, y') + o(\varepsilon). \tag{52}$$

Assuming that the curve

$$y_i = y_i(x) \qquad (1 \leqslant i \leqslant n)$$

is an extremal of $J[y]$, we can use formula (11) of Sec. 13 to write an expression for the variation of $J[y]$ corresponding to the transformation (52). Since in the present case[13]

$$\delta x = \varepsilon\varphi, \qquad \delta y_i = \varepsilon\psi_i,$$

the result is

$$\delta J = \varepsilon \left[\sum_{i=1}^{n} F_{y_i'} \psi_i + \left(F - \sum_{i=1}^{n} y_i' F_{y_i'} \right) \varphi \right]_{x=x_0}^{x=x_1}.$$

---

[12] As usual, $\eta = o(\varepsilon)$ means that $\eta/\varepsilon \to 0$ as $\varepsilon \to 0$.

[13] Here δx, δy_i mean the principal linear parts (relative to ε) of the increments Δx, Δy_i of x, y_i, and not simply Δx, Δy_i as in Sec. 13. It is easy to see that this change in interpretation has no effect on the final result, and has the advantage of making it unnecessary to bother with infinitesimals of higher order.

Since by hypothesis, $J[y]$ is invariant under (52), δJ vanishes, i.e.,

$$\left[\sum_{i=1}^{n} F_{y_i'}\psi_i + \left(F - \sum_{i=1}^{n} y_i' F_{y_i'} \right) \varphi \right]_{x=x_0}$$
$$= \left[\sum_{i=1}^{n} F_{y_i'}\psi_i + \left(F - \sum_{i=1}^{n} y_i' F_{y_i'} \right) \varphi \right]_{x=x_1}$$

The fact that (50) holds along each extremal now follows from the arbitrariness of x_0 and x_1.

Remark. In terms of the canonical variables p_i and H, equation (50) becomes simply

$$\sum_{i=1}^{n} p_i\psi_i - H\varphi = \text{const.} \tag{53}$$

Example 3. Consider the functional

$$J[y] = \int_{x_0}^{x_1} F(y, y')\, dx, \tag{54}$$

whose integrand does not depend on x explicitly. Then, by exactly the same argument as given in Example 1, $J[y]$ is invariant under the one-parameter family of transformations

$$x^* = x + \varepsilon, \qquad y_i^* = y_i. \tag{55}$$

In this case,

$$\varphi = 1, \qquad \psi_i = 0,$$

and (53) reduces to just

$$H = \text{const,}$$

i.e., the Hamiltonian H is constant along each extremal of $J[y]$. Thus, we again obtain a result already proved in Sec. 17: *For a functional of the form* (54), *which does not depend on x explicitly, the Hamiltonian is a first integral of the system of Euler equations.*

21. The Principle of Least Action

We now apply the general results obtained in the preceding sections to some mechanical problems. Suppose we are given a system of n particles (mass points), where no constraints whatsoever are imposed on the system. Let the ith particle have mass m_i and coordinates x_i, y_i, z_i $(i = 1, \ldots, n)$. Then the *kinetic energy* of the system is [14]

$$T = \frac{1}{2} \sum_{i=1}^{n} m_i(\dot{x}_i^2 + \dot{y}_i^2 + \dot{z}_i^2). \tag{56}$$

---

[14] Here t denotes the time, and the overdot denotes differentiation with respect to t.

We assume that the system has *potential energy* U, i.e., that there exists a function

$$U = U(t, x_1, y_1, z_1, \ldots, x_n, y_n, z_n) \tag{57}$$

such that the force acting on the ith particle has components

$$X_i = -\frac{\partial U}{\partial x_i}, \quad Y_i = -\frac{\partial U}{\partial y_i}, \quad Z_i = -\frac{\partial U}{\partial z_i}.$$

Next, we introduce the expression

$$L = T - U, \tag{58}$$

called the *Lagrangian* (*function*) of the system of particles. Obviously, L is a function of the time t and of the positions (x_i, y_i, z_i) and velocities $(\dot{x}_i, \dot{y}_i, \dot{z}_i)$ of the n particles in the system.

Suppose that at time t_0 the system is in some fixed position. Then the subsequent evolution of the system in time is described by a curve

$$x_i = x_i(t), \quad y_i = y_i(t), \quad z_i = z_i(t) \qquad (i = 1, \ldots, n)$$

in a space of $3n$ dimensions. It can be shown that among all curves passing through the point corresponding to the initial position of the system, the curve which actually describes the motion of the given system, under the influence of the forces acting upon it, satisfies the following condition, known as the *principle of least action*:

THEOREM. *The motion of a system of n particles during the time interval $[t_0, t_1]$ is described by those functions $x_i(t), y_i(t), z_i(t), 1 \leqslant i \leqslant n$, for which the integral*

$$\int_{t_0}^{t_1} L \, dt, \tag{59}$$

called the action, is a minimum.

Proof. We show that the principle of least action implies the usual equations of motion for a system of n particles. If the functional (59) has a minimum, then the Euler equations

$$\frac{\partial L}{\partial x_i} - \frac{d}{dt} \frac{\partial L}{\partial \dot{x}_i} = 0,$$

$$\frac{\partial L}{\partial y_i} - \frac{d}{dt} \frac{\partial L}{\partial \dot{y}_i} = 0, \tag{60}$$

$$\frac{\partial L}{\partial z_i} - \frac{d}{dt} \frac{\partial L}{\partial \dot{z}_i} = 0$$

must be satisfied for $i = 1, \ldots, n$. Bearing in mind that the potential energy U depends only on t, x_i, y_i, z_i, and not on $\dot{x}_i, \dot{y}_i, \dot{z}_i$, while T is a sum of squares of the velocity components $\dot{x}_i$, $\dot{y}_i$, $\dot{z}_i$ (with coefficients $\frac{1}{2}m_i$), we can write the equations (60) in the form

$$-\frac{\partial U}{\partial x_i} - \frac{d}{dt} m_i \dot{x}_i = 0,$$
$$-\frac{\partial U}{\partial y_i} - \frac{d}{dt} m_i \dot{y}_i = 0, \qquad (61)$$
$$-\frac{\partial U}{\partial z_i} - \frac{d}{dt} m_i \dot{z}_i = 0.$$

Finally, since the derivatives

$$-\frac{\partial U}{\partial x_i}, \quad -\frac{\partial U}{\partial y_i}, \quad -\frac{\partial U}{\partial z_i}$$

are the components of the force acting on the *i*th particle, the system (61) reduces to

$$m_i \ddot{x}_i = X_i,$$
$$m_i \ddot{y}_i = Y_i,$$
$$m_i \ddot{z}_i = Z_i,$$

which are just Newton's equations of motion for a system of *n* particles, subject to no constraints.

Remark 1. The principle of least action remains valid in the case where the system of particles is subject to constraints, except that then the admissible curves, for which the functional (59) is considered, have to satisfy the constraints. In other words, in this case, application of the principle of least action leads to a variational problem with subsidiary conditions.

Remark 2. Actually, as we shall see later (Sec. 36.2), the principle of least action only holds for sufficiently small time intervals $[t_0, t_1]$, and has to be modified for continuous mechanical systems.

22. Conservation Laws

We have just seen that the equations of motion of a mechanical system consisting of *n* particles, with kinetic energy (56), potential energy (57) and Lagrangian (58), can be obtained from the principle of least action, i.e., by minimizing the integral

$$\int_{t_0}^{t_1} L \, dt = \int_{t_0}^{t_1} (T - U) \, dt. \qquad (62)$$

The canonical variables corresponding to the functional (62) turn out to be

$$p_{ix} = \frac{\partial L}{\partial \dot{x}_i} = m_i \dot{x}_i,$$
$$p_{iy} = \frac{\partial L}{\partial \dot{y}_i} = m_i \dot{y}_i,$$
$$p_{iz} = \frac{\partial L}{\partial \dot{z}_i} = m_i \dot{z}_i,$$

which are just the components of the momentum of the ith particle.[15]   In terms of p_{ix}, p_{iy} and p_{iz}, we have

$$H = \sum_{i=1}^{n} (\dot{x}_i p_{ix} + \dot{y}_i p_{iy} + \dot{z}_i p_{iz}) - L = 2T - (T - U) = T + U,$$

so that H is the *total energy* of the system.

Using the form of the integrand in (62), we can find various functions which maintain constant values along each trajectory of the system, thereby obtaining so-called *conservation laws*.

1. *Conservation of energy.* Suppose the given system is *conservative*, which means that the Lagrangian L (or more precisely, the potential energy U) does not depend on time explicitly. Then, as shown in Sec. 17 (see also Sec. 20, Example 3), $H = $ const along each extremal, i.e., the total energy of a conservative system does not change during the motion of the system.

2. *Conservation of momentum.* First, we recall that according to Noether's theorem (Sec. 20), invariance of the functional (49) under the family of transformations

$$x^* = \Phi(x, y, y'; \varepsilon) = x,$$
$$y_i^* = \Psi_i(x, y, y'; \varepsilon)$$

implies that the corresponding system of Euler equations has the first integral

$$\sum_{i=1}^{n} F_{y_i} \psi_i = \text{const},$$

where

$$\psi_i(x, y, y') = \frac{\partial \Psi_i(x, y, y'; \varepsilon)}{\partial \varepsilon}\bigg|_{\varepsilon = 0},$$

since in this case,

$$\varphi(x, y, y') = \frac{\partial \Phi(x, y, y'; \varepsilon)}{\partial \varepsilon}\bigg|_{\varepsilon = 0} = 0.$$

Therefore, the invariance of the functional (62) under the transformation

$$x_i^* = x_i + \varepsilon, \qquad y_i^* = y_i, \qquad z_i^* = z_i$$

implies that

$$\sum_{i=1}^{n} \frac{\partial L}{\partial \dot{x}_i} = \text{const},$$

i.e.,

$$\sum_{i=1}^{n} p_{ix} = \text{const}.$$

---

[15] By analogy with mechanical problems, the variables $p_i = F_{y_i}$ are often called the *momenta*, regardless of the interpretation of the integrand F appearing in the functional (1).

Similarly, it follows from the invariance of (62) under displacements along the y-axis that

$$\sum_{i=1}^{n} p_{iy} = \text{const},$$

and from the invariance of (62) under displacements along the z-axis that

$$\sum_{i=1}^{n} p_{iz} = \text{const}.$$

The vector **P** with components

$$P_x = \sum_{i=1}^{n} p_{ix}, \qquad P_y = \sum_{i=1}^{n} p_{iy}, \qquad P_z = \sum_{i=1}^{n} p_{iz}$$

is called the *total momentum* of the system. Thus, we have just proved that the total momentum is conserved during the motion of the system if the integral (62) is invariant under parallel displacements. [It is clear from these considerations that the invariance of (62) under displacements along any coordinate axis, e.g., along the x-axis, implies that the corresponding component of the total momentum is conserved.]

3. *Conservation of angular momentum.* Suppose the integral (62) is invariant under rotations about the z-axis, i.e., under coordinate transformations of the form

$$\begin{aligned} x_i^* &= x_i \cos \varepsilon + y_i \sin \varepsilon, \\ y_i^* &= -x_i \sin \varepsilon + y_i \cos \varepsilon, \\ z_i^* &= z_i. \end{aligned}$$

In this case,

$$\begin{aligned} \psi_{ix} &= \frac{\partial x_i^*}{\partial \varepsilon}\bigg|_{\varepsilon=0} = y_i, \\ \psi_{iy} &= \frac{\partial y_i^*}{\partial \varepsilon}\bigg|_{\varepsilon=0} = -x_i, \\ \psi_{iz} &= \frac{\partial z_i^*}{\partial \varepsilon}\bigg|_{\varepsilon=0} = 0, \end{aligned}$$

and hence Noether's theorem implies that

$$\sum_{i=1}^{n} \left(\frac{\partial L}{\partial \dot{x}_i} y_i - \frac{\partial L}{\partial \dot{y}_i} x_i \right) = \text{const},$$

i.e.,

$$\sum_{i=1}^{n} (p_{ix} y_i - p_{iy} x_i) = \text{const}. \tag{63}$$

Each term in this sum represents the z-component of the vector product $\mathbf{p}_i \times \mathbf{r}_i$, where $\mathbf{r}_i = (x_i, y_i, z_i)$ is the position vector and $\mathbf{p}_i = (p_{ix}, p_{iy}, p_{iz})$ the momentum of the ith particle. The vector $\mathbf{p}_i \times \mathbf{r}_i$ is called the *angular momentum* of the ith particle, about the origin of coordinates, and (63) means that the sum of the z-components of the angular momenta of the separate particles, i.e., the z-component of the *total angular momentum* (of the whole system) is a constant. Similar assertions hold for the x and y-components of the total angular momentum, provided that the integral (62) is invariant under rotations about the x and y-axes. Thus, we have proved that the total angular momentum does not change during the motion of the system if (62) is invariant under all rotations.

Example 1. Consider the motion of a particle which is attracted to a fixed point, according to some law. In this case, energy is conserved, since L is time-invariant, and angular momentum is also conserved, since L is invariant under rotations. However, momentum is not conserved during the motion of the particle.

Example 2. A particle is attracted to a homogeneous linear mass distribution lying along the z-axis. In this case, the following quantities are conserved:

1. The energy (since L is independent of time);
2. The z-component of the momentum;
3. The z-component of the angular momentum.

23. The Hamilton-Jacobi Equation. Jacobi's Theorem[16]

Consider the functional

$$J[y] = \int_{x_0}^{x_1} F(x, y_1, \ldots, y_n, y'_1, \ldots, y'_n) \, dx \qquad (64)$$

defined on the curves lying in some region R, and suppose that one and only one extremal of (64) goes through two arbitrary points A and B. The integral

$$S = \int_{x_0}^{x_1} F(x, y_1, \ldots, y_n, y'_1, \ldots, y'_n) \, dx \qquad (65)$$

evaluated along the extremal joining the points

$$A = (x_0, y_1^0, \ldots, y_n^0), \qquad B = (x_1, y_1^1, \ldots, y_n^1) \qquad (66)$$

is called the *geodetic distance* between A and B. The quantity S is obviously a single-valued function of the coordinates of the points A and B.

---

[16] In this section, we drop the vector notation introduced in Sec. 20, and revert to the more explicit notation used earlier. The vector notation will be used again later (e.g., in Sec. 29).

Example 1. If the functional J is arc length, S is the distance (in the usual sense) between the points A and B.

Example 2. Consider the propagation of light in an inhomogeneous and anisotropic medium, where it is assumed that the velocity of light at any point depends both on the coordinates of the point and on the direction of propagation, i.e.,

$$v = v(x, y, z, \dot{x}, \dot{y}, \dot{z}).$$

The time it takes light to go from one point to another along some curve

$$x = x(t), \qquad y = y(t), \qquad z = z(t)$$

is given by the integral

$$T = \int_{t_0}^{t_1} \frac{\sqrt{\dot{x}^2 + \dot{y}^2 + \dot{z}^2}}{v}\, dt. \tag{67}$$

According to Fermat's principle, light propagates in any medium along the curve for which the transit time T is smallest, i.e., along the extremal of the functional (67). Thus, for the functional (67), S is the time it takes light to go from the point A to the point B.

Example 3. Consider a mechanical system with Lagrangian L. According to Sec. 21, the integral

$$\int_{t_0}^{t_1} L(t, x_1, y_1, z_1, \ldots, x_n, y_n, z_n)\, dt$$

evaluated along the extremal passing through two given points, i.e., two configurations of the system, is the "least action" corresponding to the motion of the system from the first configuration to the second.

If the initial point A is regarded as fixed and the final point $B = (x, y_1, \ldots, y_n)$ is regarded as variable,[17] then in the region R,

$$S = S(x, y_1, \ldots, y_n) \tag{68}$$

is a single-valued function of the coordinates of the point B. We now derive a differential equation satisfied by the function (68). We first calculate the partial derivatives

$$\frac{\partial S}{\partial x}, \quad \frac{\partial S}{\partial y_i} \qquad (i = 1, \ldots, n),$$

by writing down the total differential of the function S, i.e., the principal linear part of the increment

$$\Delta S = S(x + dx, y_1 + dy_1, \ldots, y_n + dy_n) - S(x, y_1, \ldots, y_n).$$

Since, by definition, ΔS is the difference

$$J[\gamma^*] - J[\gamma],$$

---

[17] Since B is now variable, we drop the superscript in the second of the formulas (66).

where γ is the extremal going from A to the point $(x, y_1, \ldots, y_n)$ and γ^* is the extremal going from A to the point $(x + dx, y_1 + dy_1, \ldots, y_n + dy_n)$, we have

$$dS = \delta J,$$

where the "unvaried" curve is the extremal γ and the initial point A is held fixed. (The fact that the "varied" curve γ^* is also an extremal is not important here.)

Thus, using formula (12) of Sec. 13 for the general variation of a functional, we obtain

$$dS(x, y_1, \ldots, y_n) = \delta J = \sum_{i=1}^{n} p_i \, dy_i - H \, dx, \tag{69}$$

where (69) is evaluated at the point B. It follows that

$$\frac{\partial S}{\partial x} = -H, \qquad \frac{\partial S}{\partial y_i} = p_i, \tag{70}$$

where [18]

$$p_i = p_i(x, y_1, \ldots, y_n) = F_{y_i'}[x, y_1, \ldots, y_n, y_1'(x), \ldots, y_n'(x)] \tag{71}$$

and

$$H = H[x, y_1, \ldots, y_n, p_1(x, y_1, \ldots, y_n), \ldots, p_n(x, y_1, \ldots, y_n)]$$

are functions of $x, y_1, \ldots, y_n$. Then from (70) we find that S, as a function of the coordinates of the point B, satisfies the equation

$$\frac{\partial S}{\partial x} + H\left(x, y_1, \ldots, y_n, \frac{\partial S}{\partial y_1}, \ldots, \frac{\partial S}{\partial y_n}\right) = 0. \tag{72}$$

The partial differential equation (72), which is in general nonlinear, is called the *Hamilton-Jacobi equation*. There is an intimate connection between the Hamilton-Jacobi equation and the canonical Euler equations. In fact, the canonical equations represent the so-called *characteristic system* associated with equation (72).[19] We shall approach this matter from a somewhat different point of view, by establishing a connection between solutions of the Hamilton-Jacobi equation and first integrals of the system of Euler equations:

THEOREM 1. *Let*

$$S = S(x, y_1, \ldots, y_n, \alpha_1, \ldots, \alpha_m) \tag{73}$$

---

[18] In (71), $y_i'(x)$ denotes the derivative dy_i/dx calculated at the point B for the extremal γ going from A to B.

[19] See e.g., R. Courant and D. Hilbert, *Methods of Mathematical Physics, Vol. II*, Interscience, Inc., New York (1962), Chap. 2, Sec. 8.

be a solution, depending on m $(\leqslant n)$ parameters $\alpha_1, \ldots, \alpha_m$ of the Hamilton-Jacobi equation (72). Then each derivative

$$\frac{\partial S}{\partial \alpha_i} \qquad (i = 1, \ldots, m)$$

is a first integral of the system of canonical Euler equations

$$\frac{dy_i}{dx} = \frac{\partial H}{\partial p_i}, \qquad \frac{dp_i}{dx} = -\frac{\partial H}{\partial y_i},$$

i.e.,

$$\frac{\partial S}{\partial \alpha_i} = \text{const} \qquad (i = 1, \ldots, m)$$

along each extremal.

   Proof.   We have to show that

$$\frac{d}{dx}\left(\frac{\partial S}{\partial \alpha_i}\right) = 0 \qquad (i = 1, \ldots, m) \tag{74}$$

along each extremal. Calculating the left-hand side of (74), we find that

$$\frac{d}{dx}\left(\frac{\partial S}{\partial \alpha_i}\right) = \frac{\partial^2 S}{\partial x\, \partial \alpha_i} + \sum_{k=1}^{n} \frac{\partial^2 S}{\partial y_k\, \partial \alpha_i} \frac{dy_k}{dx}. \tag{75}$$

Substituting (73) into the Hamilton-Jacobi equation (72), and differentiating the result with respect to α_i, we obtain

$$\frac{\partial^2 S}{\partial x\, \partial \alpha_i} = -\sum_{k=1}^{n} \frac{\partial H}{\partial p_k} \frac{\partial^2 S}{\partial y_k\, \partial \alpha_i}. \tag{76}$$

Then substitution of (76) into (75) gives

$$\begin{aligned}
\frac{d}{dx}\left(\frac{\partial S}{\partial \alpha_i}\right) &= -\sum_{k=1}^{n} \frac{\partial H}{\partial p_k} \frac{\partial^2 S}{\partial y_k\, \partial \alpha_i} + \sum_{k=1}^{n} \frac{\partial^2 S}{\partial y_k\, \partial \alpha_i} \frac{dy_k}{dx} \\
&= \sum_{k=1}^{n} \frac{\partial^2 S}{\partial y_k\, \partial \alpha_i}\left(\frac{dy_k}{dx} - \frac{\partial H}{\partial p_k}\right).
\end{aligned}$$

Since

$$\frac{dy_k}{dx} - \frac{\partial H}{\partial p_k} = 0 \qquad (k = 1, \ldots, n)$$

along each extremal, it follows that (74) holds along each extremal, which proves the theorem.

   THEOREM 2 (Jacobi).   Let

$$S = S(x, y_1, \ldots, y_n, \alpha_1, \ldots, \alpha_n) \tag{77}$$

be a complete integral of the Hamilton-Jacobi equation (72), i.e., a general *solution of (72) depending on n parameters* $\alpha_1, \ldots, \alpha_n$. *Moreover, let the determinant of the* $n \times n$ *matrix*

$$\left\| \frac{\partial^2 S}{\partial \alpha_i \, \partial y_k} \right\| \tag{78}$$

be nonzero, and let $\beta_1, \ldots, \beta_n$ *be* n *arbitrary constants. Then the functions*

$$y_i = y_i(x, \alpha_1, \ldots, \alpha_n, \beta_1, \ldots, \beta_n) \qquad (i = 1, \ldots, n) \tag{79}$$

defined by the relations

$$\frac{\partial}{\partial \alpha_i} S(x, y_1, \ldots, y_n, \alpha_1, \ldots, \alpha_n) = \beta_i \qquad (i = 1, \ldots, n), \tag{80}$$

together with the functions

$$p_i = \frac{\partial}{\partial y_i} S(x, y_1, \ldots, y_n, \alpha_1, \ldots, \alpha_n) \qquad (i = 1, \ldots, n), \tag{81}$$

where the y_i *are given by (79), constitute a general solution of the canonical system*

$$\frac{dy_i}{dx} = \frac{\partial H}{\partial p_i}, \qquad \frac{dp_i}{dx} = -\frac{\partial H}{\partial y_i} \qquad (i = 1, \ldots, n). \tag{82}$$

Proof 1. According to Theorem 1, the n relations (80) correspond to first integrals of the canonical system (82). To obtain the general solution of (82), we first use (80) to define the n functions (79) [this is possible since (78) has a nonvanishing determinant], and then use (81) to define the n functions p_i. To show that the functions y_i and p_i so defined actually satisfy the canonical equations (82), we argue as follows: Differentiating (80) with respect to x, where the y_i are regarded as functions of x [cf. (79)], we obtain

$$\frac{d}{dx}\left(\frac{\partial S}{\partial \alpha_i}\right) = \frac{\partial^2 S}{\partial x \, \partial \alpha_i} + \sum_{k=1}^{n} \frac{\partial^2 S}{\partial y_k \, \partial \alpha_i} \frac{dy_k}{dx} = \sum_{k=1}^{n} \frac{\partial^2 S}{\partial y_k \, \partial \alpha_i} \left(\frac{dy_k}{dx} - \frac{\partial H}{\partial p_k}\right),$$

where in the last step we have used (76). Since the determinant of the matrix (78) is nonzero, it follows that

$$\frac{dy_i}{dx} = \frac{\partial H}{\partial p_i} \qquad (i = 1, \ldots, n), \tag{83}$$

which is just the first set of equations (82).

Next, we differentiate (81) with respect to x, obtaining

$$\frac{dp_i}{dx} = \frac{d}{dx}\left(\frac{\partial S}{\partial y_i}\right) = \frac{\partial^2 S}{\partial x \, \partial y_i} + \sum_{k=1}^{n} \frac{\partial^2 S}{\partial y_k \, \partial y_i} \frac{dy_k}{dx} = \frac{\partial^2 S}{\partial x \, \partial y_i} + \sum_{k=1}^{n} \frac{\partial^2 S}{\partial y_k \, \partial y_i} \frac{\partial H}{\partial p_k},$$

where we have used (83). Then, taking account of (81) and differentiating the Hamilton-Jacobi equation (72) with respect to y_i, we find that

$$\frac{\partial^2 S}{\partial x \, \partial y_i} = -\frac{\partial H}{\partial y_i} - \sum_{k=1}^{n} \frac{\partial H}{\partial p_k} \frac{\partial^2 S}{\partial y_k \, \partial y_i}.$$

A comparison of the last two equations shows that

$$\frac{dp_i}{dx} = -\frac{\partial H}{\partial y_i} \qquad (i = 1, \ldots, n),$$

which is just the second set of equations (82).

Proof 2. Our second proof of Jacobi's theorem is based on the use of a canonical transformation. Let (77) be a complete integral of the Hamilton-Jacobi equation. We make a canonical transformation of the system (82), choosing the function (77) as the generating function, $\alpha_1, \ldots, \alpha_n$ as the new momenta (cf. footnote 15, p. 86), and $\beta_1, \ldots, \beta_n$ as the new coordinates. Then, according to formula (41) of Sec. 19,

$$p_i = \frac{\partial S}{\partial y_i}, \qquad \beta_i = \frac{\partial S}{\partial \alpha_i}, \qquad H^* = H + \frac{\partial S}{\partial x}.$$

But since the function S satisfies the Hamilton-Jacobi equation, we have

$$H^* = H + \frac{\partial S}{\partial x} = 0.$$

Therefore, in the new variables, the canonical equations become

$$\frac{d\alpha_i}{dx} = 0, \qquad \frac{d\beta_i}{dx} = 0,$$

from which it follows that $\alpha_i = \text{const}$, $\beta_i = \text{const}$ along each extremal. Thus, we again obtain the same n first integrals

$$\frac{\partial S}{\partial \alpha_i} = \beta_i$$

of the system of Euler equations. If we now use these equations to determine the functions (79) of the $2n$ parameters $\alpha_1, \ldots, \alpha_n, \beta_1, \ldots, \beta_n$, and if, as before, we set

$$p_i = \frac{\partial}{\partial y_i} S(x, y_1, \ldots, y_n, \alpha_1, \ldots, \alpha_n),$$

where the y_i are given by (79), we obtain $2n$ functions

$$y_i(x, \alpha_1, \ldots, \alpha_n, \beta_1, \ldots, \beta_n),$$
$$p_i(x, \alpha_1, \ldots, \alpha_n, \beta_1, \ldots, \beta_n),$$

which constitute a general solution of the canonical system (82).

PROBLEMS

1. Use the canonical Euler equations to find the extremals of the functional

$$\int \sqrt{x^2 + y^2} \sqrt{1 + y'^2} \, dx,$$

and verify that they agree with those found in Chap. 1, Prob. 22.

Hint. The Hamiltonian is

$$H(x, y, p) = -\sqrt{x^2 + y^2 - p^2},$$

and the corresponding canonical system

$$\frac{dp}{dx} = \frac{y}{\sqrt{x^2 + y^2 - p^2}}, \qquad \frac{dy}{dx} = \frac{p}{\sqrt{x^2 + y^2 - p^2}}$$

has the first integral

$$p^2 - y^2 = C^2,$$

where C is a constant.

2. Consider the action functional

$$J[x] = \frac{1}{2} \int_{t_0}^{t_1} (m\dot{x}^2 - \varkappa x^2) \, dt$$

corresponding to a *simple harmonic oscillator*, i.e., a particle of mass m acted upon by a restoring force $-\varkappa x$ (cf. Sec. 36.2). Write the canonical system of Euler equations corresponding to $J[x]$, and interpret them. Calculate the Poisson brackets $[x, p]$, $[x, H]$ and $[p, H]$. Is p a first integral of the canonical Euler equations?

3. Use the principle of least action to give a variational formulation of the problem of the plane motion of a particle of mass m attracted to the origin of coordinates by a force inversely proportional to the square of its distance from the origin. Write the corresponding equations of motion, the Hamiltonian and the canonical system of Euler equations. Calculate the Poisson brackets $[r, p_r]$, $[\theta, p_\theta]$, $[p_r, H]$ and $[p_\theta, H]$, where

$$p_r = \frac{\partial L}{\partial \dot{r}}, \qquad p_\theta = \frac{\partial L}{\partial \dot{\theta}}.$$

Is p_θ a first integral of the canonical Euler equations?

Hint. The action functional is

$$J[r, \theta] = \int_{t_0}^{t_1} \left[\frac{m}{2} (\dot{r}^2 + r^2\dot{\theta}^2) + \frac{k}{r} \right] dt,$$

where k is a constant, and r, θ are the polar coordinates of the particle.

4. Verify that the change of variables

$$Y_i = p_i, \qquad P_i = y_i$$

is a canonical transformation, and find the corresponding generating function.

5. Verify that the functional $J[r, \theta]$ of Prob. 3 is invariant under rotations, and use Noether's theorem (in polar coordinates) to find the corresponding conservation law.   What geometric fact does this law express?

Ans.   The line segment joining the particle to the origin sweeps out equal areas in equal times.

6. Write and solve the Hamilton-Jacobi equation corresponding to the functional

$$J[y] = \int_{x_0}^{x_1} y'^2 \, dx,$$

and use the result to determine the extremals of $J[y]$.

Ans.   The Hamilton-Jacobi equation is

$$4 \frac{\partial S}{\partial x} + \left(\frac{\partial S}{\partial y} \right)^2 = 0.$$

7. Write and solve the Hamilton-Jacobi equation corresponding to the functional

$$J[y] = \int_{x_0}^{x_1} f(y) \sqrt{1 + y'^2} \, dx,$$

and use the result to find the extremals of $J[y]$.

Ans.   The Hamilton-Jacobi equation is

$$\left(\frac{\partial S}{\partial x} \right)^2 + \left(\frac{\partial S}{\partial y} \right)^2 = f^2(y),$$

with solution

$$S = \alpha x + \int_{y_0}^{y} \sqrt{f^2(\eta) - \alpha^2} \, d\eta + \beta.$$

The extremals are

$$x - \alpha \int_{y_0}^{y} \frac{d\eta}{\sqrt{f^2(\eta) - \alpha^2}} = \text{const.}$$

8. Use the Hamilton-Jacobi equation to find the extremals of the functional of Prob. 1.

Hint.   Try a solution of the form $S = \frac{1}{2}(Ax^2 + 2Bxy + Cy^2)$.

9. What functional leads to the Hamilton-Jacobi equation

$$\left(\frac{\partial S}{\partial x} \right)^2 + \left(\frac{\partial S}{\partial y} \right)^2 = 1?$$

10. Prove that the Hamilton-Jacobi equation can be solved by quadratures if it can be written in the form

$$\Phi\left(x, \frac{\partial S}{\partial x} \right) + \Psi\left(y, \frac{\partial S}{\partial y} \right) = 0.$$

11. By a *Liouville surface* is meant a surface on which the arc-length functional has the form

$$J[y] = \int_{x_0}^{x_1} \sqrt{\varphi_1(x) + \varphi_2(y)} \sqrt{1 + y'^2} \, dx.$$

Prove that the equations of the geodesics on such a surface are

$$\int \frac{dx}{\sqrt{\varphi_1(x) - \alpha}} - \int \frac{dy}{\sqrt{\varphi_2(y) + \alpha}} = \beta,$$

where α and β are constants.   Show that surfaces of revolution are Liouville surfaces.

5

THE SECOND VARIATION.
SUFFICIENT CONDITIONS
FOR A WEAK EXTREMUM

Until now, in studying extrema of functionals, we have only considered a particular *necessary* condition for a functional to have a weak (relative) extremum for a given curve γ, i.e., the condition that the variation of the functional vanish for the curve γ. In this chapter, we shall derive *sufficient* conditions for a functional to have a weak extremum. To find these sufficient conditions, we must first introduce a new concept, namely, the *second variation* of a functional. We then study the properties of the second variation, and at the same time, we derive some new necessary conditions for an extremum.

As will soon be apparent, there exist sufficient conditions for an extremum which resemble the necessary conditions and are easy to apply. These sufficient conditions differ from the necessary conditions (also derived in this chapter) in much the same way as the sufficient conditions $y' = 0$, $y'' > 0$ for a function of one variable to have a minimum differ from the corresponding necessary conditions $y' = 0$, $y'' \geqslant 0$.

24. Quadratic Functionals. The Second Variation of a Functional

We begin by introducing some general concepts that will be needed later. A functional $B[x, y]$ depending on two elements x and y, belonging to some normed linear space $\mathscr{R}$, is said to be *bilinear* if it is a linear functional of y for any fixed x and a linear functional of x for any fixed y (cf. p. 8).   Thus,

$$B[x + y, z] = B[x, z] + B[y, z],$$
$$B[\alpha x, y] = \alpha B[x, y],$$

and

$$B[x, y + z] = B[x, y] + B[x, z],$$
$$B[x, \alpha y] = \alpha B[x, y]$$

for any $x, y, z \in \mathscr{R}$ and any real number α.

If we set $y = x$ in a bilinear functional, we obtain an expression called a *quadratic functional*.   A quadratic functional $A[x] = B[x, x]$ is said to be *positive definite*[1] if $A[x] > 0$ for every nonzero element x.

A bilinear functional defined on a finite-dimensional space is called a *bilinear form*.   Every bilinear form $B[x, y]$ can be represented as

$$B[x, y] = \sum_{i, k=1}^{n} b_{ik}\xi_i\eta_k,$$

where $\xi_1, \ldots, \xi_n$ and $\eta_1, \ldots, \eta_n$ are the components of the "vectors" x and y relative to some basis.[2]   If we set $y = x$ in this expression, we obtain a *quadratic form*

$$A[x] = B[x, y] = \sum_{i, k=1}^{n} b_{ik}\xi_i\xi_k.$$

Example 1. The expression

$$B[x, y] = \int_a^b x(t)y(t)\, dt$$

is a bilinear functional defined on the space $\mathscr{C}$ of all functions which are continuous in the interval $a \leqslant t \leqslant b$.   The corresponding quadratic functional is

$$A[x] = \int_a^b x^2(t)\, dt.$$

Example 2. A more general bilinear functional defined on $\mathscr{C}$ is

$$B[x, y] = \int_a^b \alpha(t)x(t)y(t)\, dt,$$

where $\alpha(t)$ is a fixed function.   If $\alpha(t) > 0$ for all t in $[a, b]$, then the corresponding quadratic functional

$$A[x] = \int_a^b \alpha(t)x^2(t)\, dt$$

is positive definite.

---

[1] Actually, the word "definite" is redundant here, but will be retained for traditional reasons.   Quadratic functionals $A[x]$ such that $A[x] \geqslant 0$ for all x will simply be called *nonnegative* (see p. 103 ff.).

[2] See e.g., G. E. Shilov, *op. cit.*, p. 114.

Example 3. The expression

$$A[x] = \int_a^b [\alpha(t)x^2(t) + \beta(t)x(t)x'(t) + \gamma(t)x'^2(t)]\, dt$$

is a quadratic functional defined on the space $\mathscr{D}_1$ of all functions which are continuously differentiable in the interval $[a, b]$.

Example 4. The integral

$$B[x, y] = \int_a^b \int_a^b K(s, t)x(s)y(t)\, ds\, dt,$$

where $K(s, t)$ is a fixed function of two variables, is a bilinear functional defined on $\mathscr{C}$. Replacing $y(t)$ by $x(t)$, we obtain a quadratic functional.

We now introduce the concept of the *second variation* (or *second differential*) of a functional. Let $J[y]$ be a functional defined on some normed linear space $\mathscr{R}$. In Chapter 1, we called the functional $J[y]$ *differentiable* if its increment

$$\Delta J[h] = J[y + h] - J[y]$$

can be written in the form

$$\Delta J[h] = \varphi[h] + \varepsilon \|h\|,$$

where $\varphi[h]$ is a linear functional and $\varepsilon \to 0$ as $\|h\| \to 0$. The quantity $\varphi[h]$ is the principal linear part of the increment $\Delta J[h]$, and is called the (*first*) *variation* [or (*first*) *differential*] of $J[y]$, denoted by $\delta J[h]$.

Similarly, we say that the functional $J[y]$ is *twice differentiable* if its increment can be written in the form

$$\Delta J[h] = \varphi_1[h] + \varphi_2[h] + \varepsilon \|h\|^2,$$

where $\varphi_1[h]$ is a linear functional (in fact, the first variation), $\varphi_2[h]$ is a quadratic functional, and $\varepsilon \to 0$ as $\|h\| \to 0$. The quadratic functional $\varphi_2[h]$ is called the *second variation* (or *second differential*) of the functional $J[y]$, and is denoted by $\delta^2 J[h]$.[3] From now on, it will be tacitly assumed that we are dealing with functionals which are twice differentiable. The second variation of such a functional is uniquely defined. This is proved in just the same way as the uniqueness of the first variation of a differentiable function (see Theorem 1 of Sec. 3.2).

THEOREM 1. *A necessary condition for the functional $J[y]$ to have a minimum for $y = \hat{y}$ is that*

$$\delta^2 J[y] \geqslant 0 \tag{1}$$

for $y = \hat{y}$ and all admissible h. For a maximum, the sign $\geqslant$ in (1) is replaced by $\leqslant$.

---

[3] The comment made in footnote 6, p. 12 applies here as well.

Proof. By definition, we have

$$\Delta J[h] = \delta J[h] + \delta^2 J[h] + \varepsilon \|h\|^2, \tag{2}$$

where $\varepsilon \to 0$ as $\|h\| \to 0$. According to Theorem 2 of Sec. 3.2, $\delta J[h] = 0$ for $y = \hat{y}$ and all admissible h, and hence (2) becomes

$$\Delta J[h] = \delta^2 J[h] + \varepsilon \|h\|^2. \tag{3}$$

Thus, for sufficiently small $\|h\|$, the sign of $\Delta J[h]$ will be the same as the sign of $\delta^2 J[h]$. Now suppose that $\delta^2 J[h_0] < 0$ for some admissible h_0. Then for any $\alpha \neq 0$, no matter how small, we have

$$\delta^2 J[\alpha h_0] = \alpha^2 \delta^2 J[h_0] < 0.$$

Hence, (3) can be made negative for arbitrarily small $\|h\|$. But this is impossible, since by hypothesis $J[y]$ has a minimum for $y = \hat{y}$, i.e.,

$$\Delta J[h] = J[\hat{y} + h] - J[\hat{y}] \geqslant 0$$

for all sufficiently small $\|h\|$. This contradiction proves the theorem.

The condition $\delta^2 J[h] \geqslant 0$ is necessary but of course not sufficient for the functional $J[y]$ to have a minimum for a given function. To obtain a sufficient condition, we introduce the following concept: We say that a quadratic functional $\varphi_2[h]$ defined on some normed linear space $\mathscr{R}$ is *strongly positive* if there exists a constant $k > 0$ such that

$$\varphi_2[h] \geqslant k \|h\|^2$$

for all h.[4]

THEOREM 2. *A sufficient condition for a functional $J[y]$ to have a minimum for $y = \hat{y}$, given that the first variation $\delta J[h]$ vanishes for $y = \hat{y}$, is that its second variation $\delta^2 J[h]$ be strongly positive for $y = \hat{y}$.*

Proof. For $y = \hat{y}$, we have $\delta J[h] = 0$ for all admissible h, and hence

$$\Delta J[h] = \delta^2 J[h] + \varepsilon \|h\|^2,$$

where $\varepsilon \to 0$ as $\|h\| \to 0$. Moreover, for $y = \hat{y}$,

$$\delta^2 J[h] \geqslant k \|h\|^2,$$

where $k = \text{const} > 0$. Thus, for sufficiently small ε_1, $|\varepsilon| < \frac{1}{2}k$ if $\|h\| < \varepsilon_1$. It follows that

$$\Delta J[h] = \delta^2 J[h] + \varepsilon \|h\|^2 > \tfrac{1}{2}k \|h\|^2 > 0$$

if $\|h\| < \varepsilon_1$, i.e., $J[y]$ has a minimum for $y = \hat{y}$, as asserted.

---

[4] In a finite-dimensional space, strong positivity of a quadratic form is equivalent to positive definiteness of the quadratic form. Therefore, a function of a finite number of variables has a minimum at a point P where its first differential vanishes, if its second differential is positive at P. In the general case, however, strong positivity is a stronger condition than positive definiteness.

25. The Formula for the Second Variation.   Legendre's Condition

Let $F(x, y, z)$ be a function with continuous partial derivatives up to order three with respect to all its arguments. (Henceforth, similar smoothness requirements will be assumed to hold whenever needed.) We now find an expression for the second variation in the case of the simplest variational problem, i.e., for functionals of the form

$$J[y] = \int_a^b F(x, y, y') \, dx, \tag{4}$$

defined for curves $y = y(x)$ with fixed end points

$$y(a) = A, \qquad y(b) = B.$$

First, we give the function $y(x)$ an increment $h(x)$ satisfying the boundary conditions

$$h(a) = 0, \qquad h(b) = 0. \tag{5}$$

Then, using Taylor's theorem with remainder, we write the increment of the functional $J[y]$ as

$$\Delta J[h] = J[y + h] - J[y]$$
$$= \int_a^b (F_y h + F_{y'} h') \, dx + \frac{1}{2} \int_a^b (\bar{F}_{yy} h^2 + 2\bar{F}_{yy'} hh' + \bar{F}_{y'y'} h'^2) \, dx, \tag{6}$$

where, as usual, the overbar indicates that the corresponding derivatives are evaluated along certain intermediate curves, i.e.,

$$\bar{F}_{yy} = F_{yy}(x, y + \theta h, y' + \theta h') \qquad (0 < \theta < 1),$$

and similarly for $\bar{F}_{yy'}$ and $\bar{F}_{y'y'}$.

If we replace $\bar{F}_{yy}$, $\bar{F}_{yy'}$ and $\bar{F}_{y'y'}$ by the derivatives F_{yy}, $F_{yy'}$ and $F_{y'y'}$ evaluated at the point $(x, y(x), y'(x))$, then (6) becomes

$$\Delta J[h] = \int_a^b (F_y h + F_{y'} h') \, dx + \frac{1}{2} \int_a^b (F_{yy} h^2 + 2F_{yy'} hh' + F_{y'y'} h'^2) \, dx + \varepsilon, \tag{7}$$

where ε can be written as

$$\int_a^b (\varepsilon_1 h^2 + \varepsilon_2 hh' + \varepsilon_3 h'^2) \, dx. \tag{8}$$

Because of the continuity of the derivatives F_{yy}, $F_{yy'}$ and $F_{y'y'}$, it follows that $\varepsilon_1, \varepsilon_2, \varepsilon_3 \to 0$ as $\|h\|_1 \to 0$, from which it is apparent that ε is an infinitesimal of order higher than 2 relative to $\|h\|_1^2$. The first term in the right-hand side of (7) is $\delta J[h]$, and the second term, which is quadratic in h, is the second variation $\delta^2 J[h]$. Thus, for the functional (4) we have

$$\delta^2 J[h] = \frac{1}{2} \int_a^b (F_{yy} h^2 + 2F_{yy'} hh' + F_{y'y'} h'^2) \, dx. \tag{9}$$

We now transform (9) into a more convenient form.   Integrating by parts and taking account of (5), we obtain

$$\int_a^b 2F_{yy'} hh' \, dx = - \int_a^b \left(\frac{d}{dx} F_{yy'} \right) h^2 \, dx.$$

Therefore, (9) can be written as

$$\delta^2 J[h] = \int_a^b (Ph'^2 + Qh^2) \, dx, \tag{10}$$

where

$$P = P(x) = \frac{1}{2} F_{y'y'}, \qquad Q = Q(x) = \frac{1}{2} \left(F_{yy'} - \frac{d}{dx} F_{yy'} \right). \tag{11}$$

This is the expression for the second variation which will be used below.

The following consequence of formulas (7) and (8) should be noted.   If $J[y]$ has an extremum for the curve $y = y(x)$, and if $y = y(x) + h(x)$ is an admissible curve, then

$$\Delta J[h] = \int_a^b (Ph'^2 + Qh^2) \, dx + \int_a^b (\xi h^2 + \eta h'^2) \, dx, \tag{12}$$

where $\xi, \eta \to 0$ as $\|h\|_1 \to 0$.   In fact, since $J[y]$ has an extremum for $y = y(x)$, the linear terms in the right-hand side of (7) vanish, while the quantity (8) can be written in the form

$$\int_a^b (\xi h^2 + \eta h'^2) \, dx$$

by integrating the term $\varepsilon_2 hh'$ by parts and using the boundary conditions (5). Formula (12) will be used later, when we derive sufficient conditions for a weak extremum (see Sec. 27).

It was proved in Sec. 24 that a necessary condition for a functional $J[y]$ to have a minimum is that its second variation $\delta^2 J[h]$ be nonnegative.   In the case of a functional of the form (4), we can use formula (10) to establish a necessary condition for the second variation to be nonnegative.   The argument goes as follows: Consider the quadratic functional (10) for functions $h(x)$ satisfying the condition $h(a) = 0$.   With this condition, the function $h(x)$ will be small in the interval $[a, b]$ if its derivative $h'(x)$ is small in $[a, b]$. However, the converse is not true, i.e., we can construct a function $h(x)$ which is itself small but has a large derivative $h'(x)$ in $[a, b]$.   This implies that the term Ph'^2 plays the dominant role in the quadratic functional (10), in the sense that Ph'^2 can be much larger than the second term Qh^2 but it cannot be much smaller than Qh^2 (it is assumed that $P \neq 0$).   Therefore, it might be expected that the coefficient $P(x)$ determines whether the functional (10) takes values with just one sign or values with both signs.   We now make this qualitative argument precise:

LEMMA. *A necessary condition for the quadratic functional*

$$\delta^2 J[h] = \int_a^b (Ph'^2 + Qh^2)\, dx, \tag{13}$$

defined for all functions $h(x) \in \mathscr{D}_1(a, b)$ *such that* $h(a) = h(b) = 0$, *to be nonnegative is that*

$$P(x) \geqslant 0 \qquad (a \leqslant x \leqslant b). \tag{14}$$

Proof. Suppose (14) does not hold, i.e., suppose (without loss of generality) that $P(x_0) = -2\beta$ ($\beta > 0$) at some point x_0 in $[a, b]$. Then, since $P(x)$ is continuous, there exists an $\alpha > 0$ such that $a \leqslant x_0 - \alpha$, $x_0 + \alpha \leqslant b$, and

$$P(x_0) < -\beta \qquad (x_0 - \alpha \leqslant x \leqslant x_0 + \alpha).$$

We now construct a function $h(x) \in \mathscr{D}_1(a, b)$ such that the functional (13) is negative. In fact, let

$$h(x) = \begin{cases} \sin^2 \dfrac{\pi (x - x_0)}{\alpha} & \text{for } x_0 - \alpha \leqslant x \leqslant x_0 + \alpha, \\ 0 & \text{otherwise.} \end{cases} \tag{15}$$

Then we have

$$\int_a^b (Ph'^2 + Qh^2)\, dx = \int_{x_0 - \alpha}^{x_0 + \alpha} P \frac{\pi^2}{\alpha^2} \sin^2 \frac{2\pi (x - x_0)}{\alpha}\, dx \tag{16}$$

$$+ \int_{x_0 - \alpha}^{x_0 + \alpha} Q \sin^4 \frac{\pi (x - x_0)}{\alpha}\, dx < -\frac{2\beta\pi^2}{\alpha} + 2M\alpha,$$

where

$$M = \max_{a \leqslant x \leqslant b} |Q(x)|.$$

For sufficiently small α, the right-hand side of (16) becomes negative, and hence (13) is negative for the corresponding function $h(x)$ defined by (15). This proves the lemma.

Using the lemma and the necessary condition for a minimum proved in Sec. 24, we immediately obtain

THEOREM (*Legendre*). *A necessary condition for the functional*

$$J[y] = \int_a^b F(x, y, y')\, dx, \qquad y(a) = A, \qquad y(b) = B$$

to have a minimum for the curve $y = y(x)$ *is that the inequality*

$$F_{y'y'} \geqslant 0$$

(*Legendre's condition*) *be satisfied at every point of the curve.*

Legendre attempted (unsuccessfully) to show that a sufficient condition for $J[y]$ to have a (weak) minimum for the curve $y = y(x)$ is that the strict inequality

$$F_{y'y'} > 0 \tag{17}$$

(the *strengthened Legendre condition*) be satisfied at every point of the curve. His approach was to first write the second variation (10) in the form

$$\delta^2 J[h] = \int_a^b [Ph'^2 + 2whh' + (Q + w')h^2]\, dx, \tag{18}$$

where $w(x)$ is an arbitrary differentiable function, using the fact that

$$0 = \int_a^b \frac{d}{dx}(wh^2)\, dx = \int_a^b (w'h^2 + 2whh')\, dx, \tag{19}$$

since $h(a) = h(b) = 0$. Next, he observed that the condition (17) would indeed be sufficient if it were possible to find a function $w(x)$ for which the integrand in (18) is a perfect square. However, this is not always possible, as was first shown by Legendre himself, since then $w(x)$ would have to satisfy the equation

$$P(Q + w') = w^2, \tag{20}$$

and although this equation is "locally solvable," it may not have a solution in a sufficiently large interval.[5]

Actually, the following argument shows that the requirement that

$$F_{y'y'}[x, y(x), y'(x)] > 0 \tag{21}$$

be satisfied at every point of an extremal $y = y(x)$ cannot be a sufficient condition for the extremal to be a minimum of the functional $J[y]$. The condition (21), like the condition

$$F_y - \frac{d}{dx} F_{y'} = 0$$

characterizing the extremal is of a "local" character, i.e., it does not pertain to the curve as a whole, but only to individual points of the curve. Therefore, if the condition (21) holds for any two curves AB and BC, it also holds for the curve AC formed by joining AB and BC. On the other hand, the fact that a functional has an extremum for each part AB and BC of some curve AC does not imply that it has an extremum for the whole curve AC. For example, a great circle arc on a given sphere is the shortest curve joining its end points if the arc consists of less than half a circle, but it is not the shortest curve (even in the class of neighboring curves) if the arc consists of more than half a circle. However, every great circle arc on a given sphere is an extremal of the functional which represents arc length on the sphere, and in fact it is easily verified that for this functional, (21) holds at every point of the great circle arc. Therefore, (21) cannot be a sufficient condition

---

[5] For example, if $P = -1$, $Q = 1$, we obtain the equation $w' + 1 + w^2 = 0$, so that $w(x) = \tan(c - x)$. If $b - a > \pi$, there is no solution in the whole interval $[a, b]$, since then $\tan(c - x)$ must become infinite somewhere in $[a, b]$.

for an extremum, nor, for that matter, can any set of purely local conditions be sufficient.

Although the condition (20) does not guarantee a minimum, the idea of completing the square of the integrand in formula (18) for the second variation, with the aim of finding sufficient conditions for an extremum, turns out to be very fruitful. In fact, the differential equation (20), which comes to the fore when trying to implement this idea, leads to new necessary conditions for an extremum (which are no longer local!). We shall discuss these matters further in the next two sections.

26. Analysis of the Quadratic Functional $\int_a^b (Ph'^2 + Qh^2)\, dx$

As shown in the preceding section, to pursue our study of the "simplest" variational problem, i.e., that of finding the extrema of the functional

$$J[y] = \int_a^b F(x, y, y')\, dx, \tag{22}$$

where

$$y(a) = A, \qquad y(b) = B,$$

we have to analyze the quadratic functional[6]

$$\int_a^b (Ph'^2 + Qh^2)\, dx, \tag{23}$$

defined on the set of functions $h(x)$ satisfying the conditions

$$h(a) = 0, \qquad h(b) = 0. \tag{24}$$

Here, the functions P and Q are related to the function F appearing in the integrand of (22) by the formulas

$$P = \frac{1}{2} F_{y'y'}, \qquad Q = \frac{1}{2}\left(F_{yy} - \frac{d}{dx} F_{yy'}\right). \tag{25}$$

For the time being, we ignore the fact that (23) is a second variation, satisfying the relations (25), and instead, we treat the analysis of (23) as an independent problem, in its own right.

In the last section, we saw that the condition

$$P(x) \geqslant 0 \qquad (a \leqslant x \leqslant b)$$

is necessary but not sufficient for the quadratic functional (23) to be $\geqslant 0$ for all admissible $h(x)$. In this section, it will be assumed that the strengthened inequality

$$P(x) > 0 \qquad (a \leqslant x \leqslant b)$$

---

[6] Similarly, the study of extrema of functions of several variables (in particular, the derivation of sufficient conditions for an extremum) involves the analysis of a quadratic form (the second differential).

holds.   We then proceed to find conditions which are both necessary and sufficient for the functional (23) to be >0 for all admissible $h(x) \not\equiv 0$, i.e., to be *positive definite*.   We begin by writing the Euler equation

$$-\frac{d}{dx}(Ph') + Qh = 0 \tag{26}$$

corresponding to the functional (23).[7]   This is a linear differential equation of the second order, which is satisfied, together with the boundary conditions (24), or more generally, the boundary conditions

$$h(a) = 0, \qquad h(c) = 0, \qquad (a < c \leqslant b),$$

by the function $h(x) \equiv 0$.   However, in general, (26) can have other, nontrivial solutions satisfying the same boundary conditions.   In this connection, we introduce the following important concept:

DEFINITION.   *The point $\tilde{a}$ ($\neq a$) is said to be conjugate to the point a if the equation (26) has a solution which vanishes for $x = a$ and $x = \tilde{a}$ but is not identically zero.*

Remark.   If $h(x)$ is a solution of (26) which is not identically zero and satisfies the conditions $h(a) = h(c) = 0$, then $Ch(x)$ is also such a solution, where $C = \text{const} \neq 0$.   Therefore, for definiteness, we can impose some kind of normalization on $h(x)$, and in fact we shall usually assume that the constant C has been chosen to make $h'(a) = 1$.[8]

The following theorem effectively realizes Legendre's idea, mentioned on p. 104.

THEOREM 1.   *If*

$$P(x) > 0 \qquad (a \leqslant x \leqslant b),$$

and if the interval $[a, b]$ contains no points conjugate to a, then the quadratic functional

$$\int_a^b (Ph'^2 + Qh^2)\, dx \tag{27}$$

is positive definite for all $h(x)$ such that $h(a) = h(b) = 0$.

---

[7] It must not be thought that this is done in order to find the minimum of the functional (23).   In fact, because of the homogeneity of (23), its minimum is either 0 if the functional is positive definite, or $-\infty$ otherwise.   In the latter case, it is obvious that the minimum cannot be found from the Euler equation.   The importance of the Euler equation (26) in our analysis of the quadratic functional (23) will become apparent in Theorem 1.   The reader should also not be confused by our use of the same symbol $h(x)$ to denote both admissible functions, in the domain of the functional (23), and solutions of equation (26).   This notation is convenient, but whereas admissible functions must satisfy $h(a) = h(b) = 0$, the condition $h(b) = 0$ will usually be explicitly precluded for nontrivial solutions of (26).

[8] If $h(x) \not\equiv 0$ and $h(a) = 0$, then $h'(a)$ must be nonzero, because of the uniqueness theorem for the linear differential equation (26).   See e.g., E. A. Coddington, *An Introduction to Ordinary Differential Equations*, Prentice-Hall, Inc., Englewood Cliffs, New Jersey (1961), pp. 105, 260.

Proof. The fact that the functional (27) is positive definite will be proved if we can reduce it to the form

$$\int_a^b P(x)\varphi^2(\cdots)\,dx,$$

where $\varphi^2(\cdots)$ is some expression which cannot be identically zero unless $h(x) \equiv 0$. To achieve this, we add a quantity of the form $d(wh^2)$ to the integrand of (27), where $w(x)$ is a differentiable function. This will not change the value of the functional (27), since $h(a) = h(b) = 0$ implies that

$$\int_a^b d(wh^2)\,dx = 0$$

[cf. equation (19)].

We now select a function $w(x)$ such that the expression

$$Ph'^2 + Qh^2 + \frac{d}{dx}(wh^2) = Ph'^2 + 2whh' + (Q + w')h^2 \qquad (28)$$

is a perfect square. This will be the case if $w(x)$ is chosen to be a solution of the equation

$$P(Q + w') = w^2 \qquad (29)$$

[cf. equation (20)]. In fact, if (29) holds, we can write (28) in the form

$$P\left(h' + \frac{w}{P}h\right)^2.$$

Thus, if (29) has a solution defined on the whole interval $[a, b]$, the quadratic functional (27) can be transformed into

$$\int_a^b P\left(h' + \frac{w}{P}h\right)^2 dx, \qquad (30)$$

and is therefore nonnegative.

Moreover, if (30) vanishes for some function $h(x)$, then obviously

$$h'(x) + \frac{w}{P}h(x) \equiv 0, \qquad (31)$$

since $P(x) > 0$ for $a \leq x \leq b$. Therefore the boundary condition $h(a) = 0$ implies $h(x) \equiv 0$, because of the uniqueness theorem for the first-order differential equation (31). It follows that the functional (30) is actually positive definite.

Thus, the proof of the theorem reduces to showing that the absence of points in $[a, b]$ which are conjugate to a guarantees that (29) has a solution defined on the whole interval $[a, b]$. Equation (29) is a *Riccati equation*, which can be reduced to a linear differential equation of the second order by making a change of variables. In fact, setting

$$w = -\frac{u'}{u} P, \qquad (32)$$

where u is a new unknown function, we obtain the equation

$$-\frac{d}{dx}(Pu') + Qu = 0 \qquad (33)$$

which is just the Euler equation (26) of the functional (27). If there are no points conjugate to a in $[a, b]$, then (33) has a solution which does not vanish anywhere in $[a, b]$,[9] and then there exists a solution of (29), given by (32), which is defined on the whole interval $[a, b]$. This completes the proof of the theorem.

Remark. The reduction of the quadratic functional (27) to the form (30) is the continuous analog of the reduction of a quadratic form to a sum of squares. The absence of points conjugate to a in the interval $[a, b]$ is the analog of the familiar criterion for a quadratic form to be positive definite. This connection will be discussed further in Sec. 30.

Next, we show that the absence of points conjugate to a in the interval $[a, b]$ is not only sufficient but also necessary for the functional (27) to be positive definite.

LEMMA. *If the function $h = h(x)$ satisfies the equation*

$$-\frac{d}{dx}(Ph') + Qh = 0$$

and the boundary conditions

$$h(a) = h(b) = 0, \qquad (34)$$

then

$$\int_a^b (Ph'^2 + Qh^2)\, dx = 0.$$

Proof. The lemma is an immediate consequence of the formula

$$0 = \int_a^b \left[-\frac{d}{dx}(Ph') + Qh \right] h\, dx = \int_a^b (Ph'^2 + Qh^2)\, dx,$$

which is obtained by integrating by parts and using (34).

---

[9] If the interval $[a, b]$ contains no points conjugate to a, then, since the solution of the differential equation (26) depends continuously on the initial conditions, the interval $[a, b]$ contains no points conjugate to $a - \varepsilon$, for some sufficiently small ε. Therefore, the solution which satisfies the initial conditions $h(a - \varepsilon) = 0$, $h'(a - \varepsilon) = 1$ does not vanish anywhere in the interval $[a, b]$. Implicit in this argument is the assumption that P does not vanish in $[a, b]$.

THEOREM 2. *If the quadratic functional*

$$\int_a^b (Ph'^2 + Qh^2)\, dx, \tag{35}$$

where

$$P(x) > 0 \qquad (a \leqslant x \leqslant b),$$

is positive definite for all $h(x)$ *such that* $h(a) = h(b) = 0$, *then the interval* [a, b] *contains no points conjugate to* a.

Proof. The idea of the proof is the following: We construct a family of positive definite functionals, depending on a parameter t, which for $t = 1$ gives the functional (35) and for $t = 0$ gives the very simple quadratic functional

$$\int_a^b h'^2\, dx,$$

for which there can certainly be no points in [a, b] conjugate to a. Then we prove that as the parameter t is varied continuously from 0 to 1, no conjugate points can appear in the interval [a, b].

Thus, consider the functional

$$\int_a^b [t(Ph'^2 + Qh^2) + (1 - t)h'^2]\, dx, \tag{36}$$

which is obviously positive definite for all t, $0 \leqslant t \leqslant 1$, since (35) is positive definite by hypothesis. The Euler equation corresponding to (36) is

$$-\frac{d}{dx}\{[tP + (1 - t)]h'\} + tQh = 0. \tag{37}$$

Let $h(x, t)$ be the solution of (37) such that $h(a, t) = 0$, $h_x(a, t) = 1$ for all $t, 0 \leqslant t \leqslant 1$. This solution is a continuous function of the parameter t, which for $t = 1$ reduces to the solution $h(x)$ of equation (26) satisfying the boundary conditions $h(a) = 0$, $h'(a) = 1$, and for $t = 0$ reduces to the solution of the equation $h'' = 0$ satisfying the same boundary conditions, i.e., the function $h = x - a$. We note that if $h(x_0, t_0) = 0$ at some point (x_0, t_0), then $h_x(x_0, t_0) \neq 0$. In fact, for any fixed t, $h(x, t)$ satisfies (37), and if the equations $h(x_0, t_0) = 0$, $h_x(x_0, t_0) = 0$ were satisfied simultaneously, we would have $h(x, t_0) = 0$ for all $x, a \leqslant x \leqslant b$, because of the uniqueness theorem for linear differential equations. But this is impossible, since $h_x(a, t) = 1$ for all $t, 0 \leqslant t \leqslant 1$.

Suppose now that the interval [a, b] contains a point $\tilde{a}$ conjugate to a, i.e., suppose that $h(x, 1)$ vanishes at some point $x = \tilde{a}$ in [a, b]. Then $\tilde{a} \neq b$, since otherwise, according to the lemma,

$$\int_a^b (Ph'^2 + Qh^2)\, dx = 0$$

for a function $h(x) \not\equiv 0$ satisfying the conditions $h(a) = h(b) = 0$, which would contradict the assumption that the functional (35) is positive definite. Therefore, the proof of the theorem reduces to showing that $[a, b]$ contains no *interior* point $\tilde{a}$ conjugate to a.

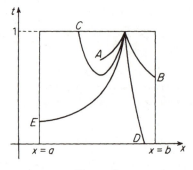

FIGURE 7

To prove this, we consider the set of all points (x, t), $a \leqslant x \leqslant b$, satisfying the condition $h(x, t) = 0$.[10] This set, if it is nonempty, represents a curve in the xt-plane, since at each point where $h(x, t) = 0$, the derivative $h_x(x, t)$ is different from zero, and hence, according to the implicit function theorem, the equation $h(x, t) = 0$ defines a continuous function $x = x(t)$ in the neighborhood of each such point.[11] By hypothesis, the point $(\tilde{a}, 1)$ lies on this curve. Thus, starting from the point $(\tilde{a}, 1)$, the curve (see Figure 7)

A. Cannot terminate inside the rectangle $a \leqslant x \leqslant b$, $0 \leqslant t \leqslant 1$, since this would contradict the continuous dependence of the solution $h(x, t)$ on the parameter t;

B. Cannot intersect the segment $x = b$, $0 \leqslant t \leqslant 1$, since then, by exactly the same argument as in the lemma [but applied to equation (37), the boundary conditions $h(a, t) = h(b, t) = 0$ and the functional (36)], this would contradict the assumption that the functional is positive definite for all t;

C. Cannot intersect the segment $a \leqslant x \leqslant b$, $t = 1$, since then for some t we would have $h(x, t) = 0$, $h_x(x, t) = 0$ simultaneously;

D. Cannot intersect the segment $a \leqslant x \leqslant b$, $t = 0$, since for $t = 0$, equation (37) reduces to $h'' = 0$, whose solution $h = x - a$ would only vanish for $x = a$;

E. Cannot approach the segment $x = a$, $0 \leqslant t \leqslant 1$, since then for some t we would have $h_x(a, t) = 0$ [why?], contrary to hypothesis.

---

[10] Recall that $h(a, t) = 0$ for all t, $0 \leqslant t \leqslant 1$.

[11] See e.g., D. V. Widder, *op. cit.*, p. 56. See also footnote 8, p. 47.

It follows that no such curve can exist, and hence the proof is complete.

If we replace the condition that the functional (35) be positive definite by the condition that it be nonnegative for all admissible $h(x)$, we obtain the following result:

THEOREM 2'. *If the quadratic functional*

$$\int_a^b (Ph'^2 + Qh^2)\, dx \tag{38}$$

where

$$P(x) > 0 \qquad (a \leqslant x \leqslant b)$$

is nonnegative for all $h(x)$ such that $h(a) = h(b) = 0$, then the interval $[a, b]$ contains no interior points conjugate to a.[12]

Proof. If the functional (38) is nonnegative, the functional (36) is positive definite for all t except possibly $t = 1$. Thus, the proof of Theorem 2 remains valid, except for the use of the lemma to prove that $\tilde{a} = b$ is impossible. Therefore, with the hypotheses of Theorem 2', the possibility that $\tilde{a} = b$ is not excluded.

Combining Theorems 1 and 2, we finally obtain

THEOREM 3. *The quadratic functional*

$$\int_a^b (Ph'^2 + Qh^2)\, dx,$$

where

$$P(x) > 0 \qquad (a \leqslant x \leqslant b),$$

is positive definite for all $h(x)$ such that $h(a) = h(b) = 0$ if and only if the interval $[a, b]$ contains no points conjugate to a.

27. Jacobi's Necessary Condition. More on Conjugate Points

We now apply the results obtained in the preceding section to the simplest variational problem, i.e., to the functional

$$\int_a^b F(x, y, y')\, dx \tag{39}$$

with the boundary conditions

$$y(a) = A, \qquad y(b) = B.$$

---

[12] In other words, the solution of the equation

$$-\frac{d}{dx}(Ph') + Qh = 0$$

satisfying the initial conditions $h(a) = 0$, $h'(a) = 1$ does not vanish at any interior point of the interval $[a, b]$.

It will be recalled from Sec. 25 that the second variation of the functional (39) [in the neighborhood of some extremal $y = y(x)$] is given by

$$\int_a^b (Ph'^2 + Qh^2)\, dx \tag{40}$$

where

$$P = \frac{1}{2} F_{y'y'}, \qquad Q = \frac{1}{2}\left(F_{yy} - \frac{d}{dx} F_{yy'}\right). \tag{41}$$

DEFINITION 1. *The Euler equation*

$$-\frac{d}{dx}(Ph') + Qh = 0 \tag{42}$$

of the quadratic functional (40) *is called the Jacobi equation of the original functional* (39).

DEFINITION 2. *The point ã is said to be conjugate to the point a with respect to the functional* (39) *if it is conjugate to a with respect to the quadratic functional* (40) *which is the second variation of* (39), *i.e., if it is conjugate to a in the sense of the definition on p. 106.*

THEOREM (*Jacobi's necessary condition*). *If the extremal $y = y(x)$ corresponds to a minimum of the junctional*

$$\int_a^b F(x, y, y')\, dx,$$

and if

$$F_{y'y'} > 0$$

along this extremal, then the open interval (a, b) contains no points conjugate to a.[13]

Proof. In Sec. 24 it was proved that nonnegativity of the second variation is a necessary condition for a minimum. Moreover, according to Theorem 2' of Sec. 26, if the quadratic functional (40) is nonnegative, the interval (a, b) can contain no points conjugate to a. The theorem follows at once from these two facts taken together.

We have just defined the Jacobi equation of the functional (39) as the Euler equation of the quadratic functional (40), which represents the second variation of (39). We can also derive Jacobi's equation by the following argument: Given that $y = y(x)$ is an extremal, let us examine the conditions which have to be imposed on $h(x)$ if the varied curve $y = y^*(x) = y(x) + h(x)$ is to be an extremal also. Substituting $y(x) + h(x)$ into Euler's equation

$$F_y(x, y + h, y' + h') - \frac{d}{dx} F_{y'}(x, y + h, y' + h') = 0,$$

---

[13] Of course, the theorem remains true if we replace the word "minimum" by "maximum" and the condition $F_{y'y'} > 0$ by $F_{y'y'} < 0$.

using Taylor's formula, and bearing in mind that $y(x)$ is already a solution of Euler's equation, we find that

$$F_{yy}h + F_{yy'}h' - \frac{d}{dx}(F_{yy'}h + F_{y'y'}h') = o(h),$$

where $o(h)$ denotes an infinitesimal of order higher than 1 relative to h and its derivative. Neglecting $o(h)$ and combining terms, we obtain the linear differential equation

$$(F_{yy} - \frac{d}{dx}F_{yy'})h - \frac{d}{dx}(F_{y'y'}h') = 0;$$

this is just Jacobi's equation, which we previously wrote in the form (42), using the notation (41). In other words, *Jacobi's equation, except for infinitesimals of order higher than 1, is the differential equation satisfied by the difference between two neighboring (i.e., "infinitely close") extremals.* An equation which is satisfied to within terms of the first order by the difference between two neighboring solutions of a given differential equation is called the *variational equation* (of the original differential equation). Thus, we have just proved that Jacobi's equation is the variational equation of Euler's equation.

Remark. These considerations are easily extended to the case of an arbitrary differential equation

$$F(x, y, y', \ldots, y^{(n)}) = 0 \qquad (43)$$

of order n. Let $y(x)$ and $y(x) + \delta y(x)$ be two neighboring solutions of (43). Replacing $y(x)$ by $y(x) + \delta y(x)$ in (43), using Taylor's formula, and bearing in mind that $y(x)$ satisfies (43), we obtain

$$F_y\delta y + F_{y'}(\delta y)' + \cdots + F_{y^{(n)}}(\delta y)^{(n)} + \varepsilon = 0,$$

where ε denotes a remainder term, which is an infinitesimal of order higher than 1 relative to δy and its derivatives. Retaining only terms of the first order, we obtain the linear differential equation

$$F_y\delta y + F_{y'}(\delta y)' + \cdots + F_{y^{(n)}}(\delta y)^{(n)} = 0,$$

satisfied by the variation δy; as before, this equation is called the *variational equation* of the original equation (43). For initial conditions which are sufficiently close to zero, this equation defines a function which is the principal linear part of the difference between two neighboring solutions of (43) with neighboring initial conditions.

We now return to the concept of a *conjugate point.* It will be recalled that in Sec. 26 the point $\tilde{a}$ was said to be conjugate to the point a if $h(\tilde{a}) = 0$, where $h(x)$ is a solution of Jacobi's equation satisfying the initial conditions $h(a) = 0$, $h'(a) = 1$. As just shown, the difference $z(x) = y^*(x) - y(x)$

corresponding to two neighboring extremals $y = y(x)$ and $y = y^*(x)$ drawn from the same initial point must satisfy the condition

$$-\frac{d}{dx}(Pz') + Qz = o(z),$$

where $o(z)$ is an infinitesimal of order higher than 1 relative to z and its derivative. Hence, to within such an infinitesimal, $y^*(x) - y(x)$ is a nonzero solution of Jacobi's equation. This leads to another definition of a conjugate point:[14]

DEFINITION 3. *Given an extremal* $y = y(x)$, *the point* $\tilde{M} = (\tilde{a}, y(\tilde{a}))$ *is said to be conjugate to the point* $M = (a, y(a))$ *if at* $\tilde{M}$ *the difference* $y^*(x) - y(x)$, *where* $y = y^*(x)$ *is any neighboring extremal drawn from the same initial point* M, *is an infinitesimal of order higher than 1 relative to* $\|y^*(x) - y(x)\|_1$.

Still another definition of a conjugate point is possible:

DEFINITION 4. *Given an extremal* $y = y(x)$, *the point* $\tilde{M} = (\tilde{a}, y(\tilde{a}))$ *is said to be conjugate to the point* $M = (a, y(a))$ *if* $\tilde{M}$ *is the limit as* $\|y^*(x) - y(x)\|_1 \to 0$ *of the points of intersection of* $y = y(x)$ *and the neighboring extremals* $y = y^*(x)$ *drawn from the same initial point* M.

It is clear that if the point $\tilde{M}$ is conjugate to the point M in the sense of Definition 4 (i.e., if the extremals intersect in the way described), then $\tilde{M}$ is also conjugate to M in the sense of Definition 3. We now verify that the converse is true, thereby establishing the equivalence of Definitions 3 and 4. Thus, let $y = y(x)$ be the extremal under consideration, satisfying the initial condition

$$y(a) = A,$$

and let $y_\alpha^*(x)$ be the extremal drawn from the same initial point $M = (a, A)$, satisfying the condition

$$y_\alpha^{*\prime}(a) - y'(a) = \alpha.$$

Then $y_\alpha^*(x)$ can be represented in the form

$$y_\alpha^*(x) = y(x) + \alpha h(x) + \varepsilon,$$

where $h(x)$ is a solution of the appropriate Jacobi equation, satisfying the conditions

$$h(a) = 0, \qquad h'(a) = 1,$$

and ε is a quantity of order higher than 1 relative to α.

Now let

$$h(\tilde{a}) = 0, \qquad \beta = \sqrt{\frac{\varepsilon}{\alpha}}.$$

---

[14] In stating this definition, we enlarge the meaning of a conjugate point to apply to points lying on an extremal and not just their abscissas. In all these considerations, it is tacitly assumed that $P = \frac{1}{2}F_{y'y'}$ has constant sign along the given extremal $y = y(x)$.

It is clear that $h'(\tilde{a}) \neq 0$, since $h(x) \not\equiv 0$. Using Taylor's formula, we can easily verify that for sufficiently small α, the expression

$$y_\alpha^*(x) - y(x) = \alpha h(x) + \varepsilon$$

takes values with different signs at the points $\tilde{a} - \beta$ and $\tilde{a} + \beta$. Since $\beta \to 0$ as $\alpha \to 0$, this means that $\tilde{M} = (\tilde{a}, y(\tilde{a}))$ is the limit as $\alpha \to 0$ of the points of intersection of the extremals $y = y_\alpha^*(x)$ and the extremal $y = y(x)$.

Example. Consider the geodesics on a sphere, i.e., the great circle arcs. Each such arc is an extremal of the functional which gives arc length on the sphere. The conjugate of any point M on the sphere is the diametrically opposite point $\tilde{M}$. In fact, given an extremal, *all* extremals with the same initial point M (and not just the neighboring extremals) intersect the given extremal at $\tilde{M}$. This property stems from the fact that a sphere has constant curvature, and is no longer true if the sphere is replaced by a "neighboring" ellipsoid (for example).

We conclude this section by summarizing the necessary conditions for an extremum found so far: If the functional

$$\int_a^b F(x, y, y')\, dx, \qquad y(a) = A, \quad y(b) = B$$

has a weak extremum for the curve $y = y(x)$, then

1. The curve $y = y(x)$ is an extremal, i.e., satisfies Euler's equation

$$F_y - \frac{d}{dx} F_{y'}$$

(see Sec. 4);

2. Along the curve $y = y(x)$, $F_{y'y'} \geqslant 0$ for a minimum and $F_{y'y'} \leqslant 0$ for a maximum (see Sec. 25);

3. The interval (a, b) contains no points conjugate to a (see Sec. 27).

28. Sufficient Conditions for a Weak Extremum

In this section, we formulate a set of conditions which is sufficient for a functional of the form

$$J[y] = \int_a^b F(x, y, y')\, dx, \qquad y(a) = A, \quad y(b) = B \tag{44}$$

to have a weak extremum for the curve $y = y(x)$. It should be noted that the sufficient conditions to be given below closely resemble the necessary conditions given at the end of the preceding section. The necessary conditions were considered separately, since each of them is necessary by itself.

However, the sufficient conditions have to be considered as a set, since the presence of an extremum is assured only if all the conditions are satisfied simultaneously.

THEOREM. *Suppose that for some admissible curve* $y = y(x)$, *the functional* (44) *satisfies the following conditions:*

1. *The curve* $y = y(x)$ *is an extremal, i.e., satisfies Euler's equation*

$$F_y - \frac{d}{dx} F_{y'} = 0;$$

2. *Along the curve* $y = y(x)$,

$$P(x) \equiv \tfrac{1}{2} F_{y'y'}[x, y(x), y'(x)] > 0$$

(*the strengthened Legendre condition*);

3. *The interval* $[a, b]$ *contains no points conjugate to the point* a (*the strengthened Jacobi condition*).[15]

Then the functional (44) *has a weak minimum for* $y = y(x)$.

Proof. If the interval $[a, b]$ contains no points conjugate to a, and if $P(x) > 0$ in $[a, b]$, then because of the continuity of the solution of Jacobi's equation and of the function $P(x)$, we can find a larger interval $[a, b + \varepsilon]$ which also contains no points conjugate to a, and such that $P(x) > 0$ in $[a, b + \varepsilon]$. Consider the quadratic functional

$$\int_a^b (Ph'^2 + Qh^2) \, dx - \alpha^2 \int_a^b h'^2 \, dx, \tag{45}$$

with the Euler equation

$$-\frac{d}{dx}[(P - \alpha^2)h'] + Qh = 0. \tag{46}$$

Since $P(x)$ is positive in $[a, b + \varepsilon]$ and hence has a positive (greatest) lower bound on this interval, and since the solution of (46) satisfying the initial conditions $h(a) = 0$, $h'(0) = 1$ depends continuously on the parameter α for all sufficiently small α, we have

1. $P(x) - \alpha^2 > 0$, $a \leqslant x \leqslant b$;

2. The solution of (46) satisfying the boundary conditions $h(a) = 0$, $h'(a) = 1$ does not vanish for $a < x \leqslant b$.

As shown in Theorem 1 of Sec. 26, these two conditions imply that the quadratic functional (45) is positive definite for all sufficiently small α. In other words, there exists a positive number $c > 0$ such that

$$\int_a^b (Ph'^2 + Qh^2) \, dx > c \int_a^b h'^2 \, dx. \tag{47}$$

---

[15] The ordinary Jacobi condition states that the *open* interval (a, b) contains no points conjugate to a. Cf. Jacobi's necessary condition, p. 112.

It is now an easy consequence of (47) that a minimum is actually achieved for the given extremal. In fact, if $y = y(x)$ is the extremal and $y = y(x) + h(x)$ is a sufficiently close neighboring curve, then, according to formula (12) of Sec. 25,

$$J[y + h] - J[y] = \int_a^b (Ph'^2 + Qh^2)\, dx + \int_a^b (\xi h^2 + \eta h'^2)\, dx, \quad (48)$$

where $\xi(x), \eta(x) \to 0$ uniformly for $a \leqslant x \leqslant b$ as $\|h\|_1 \to 0$. Moreover, using the Schwarz inequality, we have

$$h^2(x) = \left(\int_a^x h'\, dx \right)^2 \leqslant (x - a) \int_a^x h'^2\, dx \leqslant (x - a) \int_a^b h'^2\, dx,$$

i.e.,

$$\int_a^b h^2\, dx \leqslant \frac{(b - a)^2}{2} \int_a^b h'^2\, dx,$$

which implies that

$$\left| \int_a^b (\xi h^2 + \eta h'^2)\, dx \right| \leqslant \varepsilon \left(1 + \frac{(b - a)^2}{2} \right) \int_a^b h'^2\, dx \quad (49)$$

if $|\xi(x)| \leqslant \varepsilon,\ |\eta(x)| \leqslant \varepsilon$. Since $\varepsilon > 0$ can be chosen to be arbitrarily small, it follows from (47) and (49) that

$$J[y + h] - J[y] = \int_a^b (Ph'^2 + Qh^2)\, dx + \int_a^b (\xi h^2 + \eta h'^2)\, dx > 0$$

for all sufficiently small $\|h\|_1$. Therefore, the extremal $y = y(x)$ actually corresponds to a weak minimum of the functional (44), in some sufficiently small neighborhood of $y = y(x)$. This proves the theorem, thereby establishing sufficient conditions for a weak extremum in the case of the "simplest" variational problem.

29. Generalization to n Unknown Functions

The concept of a conjugate point and the related Jacobi conditions can be generalized to the case where the functional under consideration depends on n functions $y_1(x), \ldots, y_n(x)$. In this section we carry over to such functionals the definitions and results given earlier for functionals depending on a single function. To keep the notation simple, we write

$$J[y] = \int_a^b F(x, y, y')\, dx \quad (50)$$

as before, where now y denotes the n-dimensional vector $(y_1, \ldots, y_n)$ and y'

the n-dimensional vector $(y_1', \ldots, y_n')$ [cf. Sec. 20]. By the *scalar product* (y, z) of two vectors

$$y = (y_1, \ldots, y_n), \qquad z = (z_1, \ldots, z_n)$$

we mean, as usual, the quantity

$$(y, z) = y_1 z_1 + \cdots + y_n z_n.$$

Whenever the transition from the case of a single function to the case of n functions is straightforward, we shall omit details.

29.1. The second variation. The Legendre condition. If the increment $\Delta J[h]$ of the functional (50), corresponding to the change from y to $y + h$,[16] can be written in the form

$$\Delta J[h] = \varphi_1[h] + \varphi_2[h] + \varepsilon \|h\|^2,$$

where $\varphi_1[h]$ is a linear functional, $\varphi_2[h]$ is a quadratic functional, and $\varepsilon \to 0$ as $\|h\| \to 0$, then $\varphi_2[h]$ is called the *second variation* of the original functional (50) and is denoted by $\delta^2 J[h]$.[17] In the case of fixed end points, where

$$h_i(a) = h_i(b) = 0 \qquad (i = 1, \ldots, n),$$

or more concisely,

$$h(a) = h(b) = 0,$$

we easily find, applying Taylor's formula, that the second variation of (50) is given by

$$\delta^2 J[h] = \frac{1}{2} \int_a^b \left[\sum_{i,k=1}^n F_{y_i y_k} h_i h_k + 2 \sum_{i,k=1}^n F_{y_i y_k'} h_i h_k' + \sum_{i,k=1}^n F_{y_i' y_k'} h_i' h_k' \right] dx. \quad (51)$$

Introducing the matrices

$$F_{yy} = \|F_{y_i y_k}\|, \qquad F_{yy'} = \|F_{y_i y_k'}\|, \qquad F_{y'y'} = \|F_{y_i' y_k'}\|, \quad (52)$$

we can write (51) in the compact form

$$\delta^2 J[h] = \frac{1}{2} \int_a^b [(F_{yy}h, h) + 2(F_{yy'}h, h') + (F_{y'y'}h', h')] \, dx, \quad (53)$$

where each term in the integrand is the scalar product of the vector h or h' and the vector obtained by applying one of the matrices (52) to h or h'. Then, integrating by parts, we can reduce (53) to the form

$$\int_a^b [(Ph', h') + (Qh, h)] \, dx, \quad (54)$$

---

[16] The letter h denotes the vector $(h_1, \ldots, h_n)$, and $\|h\|$ means

$$\sum_{i=1}^n \max_{a \leqslant x \leqslant b} \{|h_i(x)| + |h_i'(x)|\} = \sum_{i=1}^n \|h_i\|_1.$$

[17] Obviously, $\varphi_1[h]$ is the *(first)* variation of the functional (50).

where $P = P(x)$ and $Q = Q(x)$ are the matrices

$$P = \|P_{ik}\| = \frac{1}{2} F_{y'y'}, \qquad Q = \|Q_{ik}\| = \frac{1}{2}\left(F_{yy} - \frac{d}{dx} F_{yy'}\right).$$

In deriving (54), we assume that $F_{yy'}$ is a *symmetric* matrix,[18] i.e., that $F_{y_k y'_i} = F_{y_i y'_k}$ for all $i, k = 1, \ldots, n$ (F_{yy} and $F_{y'y'}$ are automatically symmetric, because of the tacitly assumed smoothness of F). Just as in the case of one unknown function, it is easily verified that the term (Ph', h') makes the "main contribution" to the quadratic functional (54). More precisely, we have the following result:

> THEOREM 1. *A necessary condition for the quadratic functional (54) to be nonnegative for all $h(x)$ such that $h(a) = h(b) = 0$ is that the matrix P be nonnegative definite.*[19]

29.2. Investigation of the quadratic functional (54). As in Sec. 26, we can investigate the functional (54) without reference to the original functional (50), assuming, however, that P and Q are symmetric matrices. As before (see Sec. 26), we begin by writing the system of Euler equations

$$-\frac{d}{dx} \sum_{i=1}^{n} P_{ik} h'_i + \sum_{i=1}^{n} Q_{ik} h_i = 0 \qquad (k = 1, \ldots, n), \tag{55}$$

corresponding to the functional (54). The equations (55) can be written more concisely as

$$-\frac{d}{dx}(Ph') + Qh = 0, \tag{56}$$

in terms of the matrices P and Q.

> DEFINITION 1. *Let*
>
> $$\begin{aligned} h^{(1)} &= (h_{11}, h_{12}, \ldots, h_{1n}), \\ h^{(2)} &= (h_{21}, h_{22}, \ldots, h_{2n}), \\ &\quad \cdot \quad\cdot\quad\cdot\quad\cdot\quad\cdot \\ h^{(n)} &= (h_{n1}, h_{n2}, \ldots, h_{nn}) \end{aligned} \tag{57}$$
>
> *be a set of n solutions of the system (55), where the i'th solution satisfies the initial conditions*[20]
>
> $$h_{ik}(a) = 0 \qquad (k = 1, \ldots, n) \tag{58}$$
>
> *and*
>
> $$h'_{ii}(a) = 1, \qquad h'_{ik}(a) = 0 \qquad (k \neq i). \tag{59}$$

---

[18] Without this assumption, which is unnecessarily restrictive, equations (54) and (55) become more complicated, but it can be shown that Theorems 1 and 2 remain valid (H. Niemeyer, private communication).

[19] This is the appropriate multidimensional generalization of the Legendre condition (14), p. 103. The matrix $P = P(x)$ is said to be *nonnegative definite* (*positive definite*) if the quadratic form

$$\sum_{i,k=1}^{n} P_{ik}(x)h_i(x)h_k(x) \qquad (a \leqslant x \leqslant b)$$

is nonnegative (positive) for all x in $[a, b]$ and *arbitrary* $h_1(x), \ldots, h_n(x)$.

[20] Thus, the vectors $h^{(i)}(a)$ are the rows of the zero matrix of order n, and the vectors $h^{(i)'}(a)$ are the rows of the unit matrix of order n.

Then the point $\tilde{a}\ (\neq a)$ is said to be conjugate to the point a if the determinant

$$\begin{vmatrix} h_{11}(x) & h_{12}(x) & \cdots & h_{1n}(x) \\ h_{21}(x) & h_{22}(x) & \cdots & h_{2n}(x) \\ . & . & \cdots & . \\ h_{n1}(x) & h_{n2}(x) & \cdots & h_{nn}(x) \end{vmatrix} \qquad (60)$$

vanishes for $x = \tilde{a}$.

THEOREM 2.   *If P is a positive definite symmetric matrix, and if the interval $[a, b]$ contains no points conjugate to a, then the quadratic functional (54) is positive definite for all $h(x)$ such that $h(a) = h(b) = 0$.*

Proof.   The proof of this theorem follows the same plan as the proof of Theorem 1 of Sec. 26.   Let W be an arbitrary differentiable symmetric matrix.   Then

$$0 = \int_a^b \frac{d}{dx} (Wh, h)\, dx = \int_a^b (W'h, h)\, dx + 2 \int_a^b (Wh, h')\, dx$$

for every vector h satisfying the boundary conditions (58).   Therefore, we can add the expression

$$(W'h, h) + 2(Wh, h')$$

to the integrand of (54), obtaining

$$\int_a^b [(Ph', h') + 2(Wh, h') + (Qh, h) + (W'h, h)]\, dx, \qquad (61)$$

without changing the value of (54).

We now try to select a matrix W such that the integrand of (61) is a perfect square.   This will be the case if W is chosen to be a solution of the equation[21]

$$Q + W' = WP^{-1}W, \qquad (62)$$

which we call the *matrix Riccati equation* (cf. p. 108).   In fact, if we use (62), the integrand of (61) becomes

$$(Ph', h') + 2(Wh, h') + (WP^{-1}Wh, h). \qquad (63)$$

Since P is a positive definite symmetric matrix, the square root $P^{1/2}$ exists, is itself positive definite and symmetric, and has the inverse $P^{-1/2}$.   Therefore, we can write (63) as the "perfect square"

$$(P^{1/2}h' + P^{-1/2}Wh, P^{1/2}h' + P^{-1/2}Wh).$$

[Recall that if T is a symmetric matrix, $(Ty, z) = (y, Tz)$ for any vectors y and z.]   Repeating the argument given in the case of a scalar function h (see p. 107), we can show that

$$P^{1/2}h' + P^{-1/2}Wh$$

---

[21] It can be shown that this is compatible with W being symmetric, even when $F_{yy'}$ fails to be symmetric and (62) is replaced by a more general equation (H. Niemeyer, private communication).

cannot vanish for all x in $[a, b]$ unless $h \equiv 0$. It follows that if the matrix Riccati equation (62) has a solution W defined on the whole interval $[a, b]$, then, with this choice of W, the functional (61), and hence the functional (54), is positive definite.

Thus, the proof of the theorem reduces to showing that the absence of points in $[a, b]$ which are conjugate to a guarantees that (62) has a solution defined on the whole interval $[a, b]$. Making the substitution

$$W = -PU'U^{-1} \tag{64}$$

in (62), where U is a new unknown matrix [cf. (32)], we obtain the equation

$$-\frac{d}{dx}(PU') + QU = 0, \tag{65}$$

which is just the matrix form of equation (56). The solution of (65) satisfying the initial conditions

$$U(0) = \theta, \qquad U'(0) = I,$$

where θ is the zero matrix and I the unit matrix of order n, is precisely the set of solutions (57) of the system (55) which satisfy the initial conditions (58) and (59) [cf. footnote 19, p. 119]. If $[a, b]$ contains no points conjugate to a, we can show that (65) has a solution $U(x)$ whose determinant does not vanish anywhere in $[a, b]$,[22] and then there exists a solution of (62), given by (64), which is defined on the whole interval $[a, b]$. In other words, we can actually find a matrix W which converts the integrand of the functional (61) into a perfect square, in the way described. This completes the proof of the theorem.

Next we show, as in Sec. 26, that the absence of points conjugate to a in the interval $[a, b]$ is not only sufficient but also necessary for the functional (53) to be positive definite.

LEMMA. *If*

$$h(x) = (h_1(x), \ldots, h_n(x))$$

satisfies the system (55) *and the boundary conditions*

$$h(a) = h(b) = 0, \tag{66}$$

then

$$\int_a^b [(Ph', h') + (Qh, h)] \, dx = 0.$$

---

[22] The fact that $\det P$ does not vanish in $[a, b]$ is tacitly assumed, but this is guaranteed by the positive definiteness of P (cf. footnote 9, p 108).

Proof. The lemma is an immediate consequence of the formula

$$0 = \int_a^b \left(-\frac{d}{dx}(Ph') + Qh, h \right) dx = \int_a^b [(Ph', h') + (Qh, h)] \, dx,$$

which is obtained by integrating by parts and using (66).

THEOREM 3. *If the quadratic functional*

$$\int_a^b [(Ph', h') + (Qh, h)] \, dx, \tag{67}$$

where P is a positive definite symmetric matrix, is positive definite for all $h(x)$ such that $h(a) = h(b) = 0$, then the interval $[a, b]$ contains no points conjugate to a.

Proof. The proof of this theorem follows the same plan as the proof of the corresponding theorem for the case of one unknown function (Theorem 2 of Sec. 26). We consider the positive definite quadratic functional

$$\int_a^b \{t[(Ph', h') + (Qh, h)] + (1 - t)(h', h')\} \, dx. \tag{68}$$

The system of Euler equations corresponding to (68) is

$$-\frac{d}{dx} \left[t \sum_{i=1}^n P_{ik}h'_i + (1 - t)h'_k \right] + t \sum_{i=1}^n Q_{ik}h_i = 0 \quad (k = 1, \ldots, n) \tag{69}$$

[cf. (37)], which for $t = 1$ reduces to the system (55), and for $t = 0$ reduces to the system

$$h''_k = 0 \qquad (k = 1, \ldots, n).$$

Suppose the interval $[a, b]$ contains a point $\tilde{a}$ conjugate to a, i.e., suppose the determinant (60) vanishes for $x = \tilde{a}$. Then there exists a linear combination $h(x)$ of the solutions (57) which is not identically zero such that $h(\tilde{a}) = 0$. Moreover, there exists a nontrivial solution $h(x, t)$ of the system (69) which depends continuously on t and reduces to $h(x)$ for $t = 1$. It is clear that $\tilde{a} \neq b$, since otherwise, according to the lemma, the positive definite functional (67) would vanish for $h(x) \not\equiv 0$, which is impossible. The fact that $\tilde{a}$ cannot be an interior point of $[a, b]$ is proved by the same kind of argument as used in Theorem 2 of Sec. 26, for the case of a scalar function $h(x)$. Further details are left to the reader.

Suppose now that we only require that the functional (67) be nonnegative. Then, by the same argument as used to prove Theorem 2' of Sec. 26, we have

THEOREM 3'. *If the quadratic functional*

$$\int_a^b [(Ph', h') + (Qh, h)] \, dx,$$

where P is a positive definite symmetric matrix, is nonnegative for all $h(x)$ such that $h(a) = h(b) = 0$, then the interval $[a, b]$ contains no interior points conjugate to a.

Finally, combining Theorems 2 and 3, we obtain

THEOREM 4. *The quadratic functional*

$$\int_a^b [(Ph', h') + (Qh, h)] \, dx,$$

where P is a positive definite symmetric matrix, is positive definite for all $h(x)$ such that $h(a) = h(b) = 0$ if and only if the interval $[a, b]$ contains no point conjugate to a.

29.3. Jacobi's necessary condition. More on conjugate points.

We now apply the results just obtained to the original functional

$$J[y] = \int_a^b F(x, y, y') \, dx, \qquad y(a) = M_0, \quad y(b) = M_1, \tag{70}$$

where M_0 and M_1 are two fixed points, recalling that the second variation of (70) is given by

$$\int_a^b [(Ph', h') + (Qh, h)] \, dx, \tag{71}$$

where

$$P = \frac{1}{2} F_{y'y'}, \qquad Q = \frac{1}{2}\left(F_{yy} - \frac{d}{dx} F_{yy'} \right). \tag{72}$$

DEFINITION 2. *The system of Euler equations*

$$-\frac{d}{dx} \sum_{i=1}^n P_{ik} h_i' + \sum_{i=1}^n Q_{ik} h_i = 0 \qquad (k = 1, \dots, n),$$

or more concisely

$$-\frac{d}{dx}(Ph') + Qh = 0, \tag{73}$$

of the quadratic functional (71) is called the Jacobi system of the original functional (70).[23]

DEFINITION 3. *The point $\tilde{a}$ is said to be conjugate to the point a with respect to the functional (70) if it is conjugate to a with respect to the quadratic functional (71) which is the second variation of the functional (70), i.e., if it is conjugate to a in the sense of Definition 1, p. 119.*

Since nonnegativity of the second variation is a necessary condition for the functional (70) to have a minimum (see Theorem 1 of Sec. 24), Theorem 3′ immediately implies

---

[23] Equations (70)–(73) closely resemble equations (39)–(42) of Sec. 27, except that h, h' are now vectors, and P, Q are now matrices.

THEOREM 5 (*Jacobi's necessary condition*). *If the extremal*

$$y_1 = y_1(x), \ldots, y_n = y_n(x)$$

corresponds to a minimum of the functional (70), *and if the matrix*

$$F_{y'y'}[x, y(x), y'(x)]$$

is positive definite along this extremal, then the open interval (a, b) *contains no points conjugate to a.*

So far, we have said that the point $\tilde{a}$ is conjugate to a if the determinant formed from n linearly independent solutions of the Jacobi system, satisfying certain initial conditions, vanishes for $x = \tilde{a}$. As in the case $n = 1$, this basic definition is equivalent to two others, which involve only extremals of the functional (70), and not solutions of the Jacobi system:

DEFINITION 4. *Suppose n neighboring extremals*

$$y_1 = y_{i1}(x), \ldots, y_n = y_{in}(x) \qquad (i = 1, \ldots, n)$$

start from the same n-dimensional point, with directions which are close together but linearly independent. Then the point $\tilde{a}$ is said to be conjugate to the point a if the value of the determinant

$$\begin{vmatrix} y_{11}(x) & y_{12}(x) & \cdots & y_{1n}(x) \\ y_{21}(x) & y_{22}(x) & \cdots & y_{2n}(x) \\ \cdot & \cdot & \cdots & \cdot \\ y_{n1}(x) & y_{n2}(x) & \cdots & y_{nn}(x) \end{vmatrix}$$

for $x = \tilde{a}$ is an infinitesimal whose order is higher than that of its values for $a < x < \tilde{a}$.

In the next definition, we enlarge the meaning of a conjugate point to apply to points lying on extremals (cf. footnote 14, p. 114).

DEFINITION 5. *Given an extremal γ with equations*

$$y_1 = y_1(x), \ldots, y_n = y_n(x),$$

the point

$$\tilde{M} = (\tilde{a}, y_1(\tilde{a}), \ldots, y_n(\tilde{a}))$$

is said to be conjugate to the point

$$M = (a, y_1(a), \ldots, y_n(a))$$

if γ has a sequence of neighboring extremals drawn from the same initial point M, such that each neighboring extremal intersects γ and the points of intersection have $\tilde{M}$ as their limit.

The equivalence of all these definitions of a conjugate point is proved by using considerations similar to those given for the case of a single unknown function (see Sec. 27).

29.4. Sufficient conditions for a weak extremum. Theorem 2 and an argument like that used to prove the corresponding theorem of Sec. 28 (for the scalar case) imply

THEOREM 6. *Suppose that for some admissible curve* γ *with equations*

$$y_1 = y_1(x), \ldots, y_n = y_n(x),$$

the functional (70) *satisfies the following conditions:*

1. *The curve* γ *is an extremal, i.e., satisfies the system of Euler equations*

$$F_{y_i} - \frac{d}{dx} F_{y_i'} = 0 \qquad (i = 1, \ldots, n);$$

2. *Along* γ *the matrix*

$$P(x) = \tfrac{1}{2}F_{y'y'}[x, y(x), y'(x)]$$

is positive definite;

3. *The interval* $[a, b]$ *contains no points conjugate to the point* a.

Then the functional (70) *has a weak minimum for the curve* γ.

30. Connection between Jacobi's Condition and the Theory of Quadratic Forms[24]

According to Theorem 3 of Sec. 26, the quadratic functional

$$\int_a^b (Ph'^2 + Qh^2) \, dx, \tag{74}$$

where

$$P(x) > 0 \qquad (a \leqslant x \leqslant b),$$

is positive definite for all $h(x)$ such that $h(a) = h(b) = 0$ if and only if the interval $[a, b]$ contains no points conjugate to a.[25] The functional (74) is the infinite-dimensional analog of a quadratic form. Therefore, to obtain conditions for (74) to be positive definite, it is natural to start from the conditions for a quadratic form defined on an n-dimensional space to be positive definite, and then take the limit as $n \to \infty$.

This may be done as follows: By introducing the points

$$a = x_0, x_1, \ldots, x_n, x_{n+1} = b,$$

we divide the interval $[a, b]$ into $n + 1$ equal parts of length

$$\Delta x = x_{i+1} - x_i = \frac{b - a}{n + 1} \qquad (i = 0, 1, \ldots, n).$$

---

[24] Like Sec. 29, this section is written in a somewhat more concise style than the rest of the book, and can be omitted without loss of continuity.

[25] This is the *strengthened Jacobi condition* (see p. 116).

Then we consider the quadratic form

$$\sum_{i=0}^{n} \left[P_i \left(\frac{h_{i+1} - h_i}{\Delta x} \right)^2 + Q_i h_i^2 \right] \Delta x, \tag{75}$$

where P_i, Q_i and h_i are the values of the functions $P(x)$, $Q(x)$ and $h(x)$ at the point $x = x_i$. This quadratic form is a "finite-dimensional approximation" to the functional (74). Grouping similar terms and bearing in mind that

$$h_0 = h(a) = 0, \qquad h_{n+1} = h(b) = 0,$$

we can write (75) as

$$\sum_{i=1}^{n} \left[\left(Q_i \Delta x + \frac{P_{i-1} + P_i}{\Delta x} \right) h_i^2 - 2 \frac{P_{i-1}}{\Delta x} h_{i-1} h_i \right]. \tag{76}$$

In other words, the quadratic functional (74) can be approximated by a quadratic form in n variables $h_1, \ldots, h_n$, with the $n \times n$ matrix

$$\begin{Vmatrix} a_1 & b_1 & 0 & \cdots & 0 & 0 & 0 \\ b_1 & a_2 & b_2 & \cdots & 0 & 0 & 0 \\ 0 & b_2 & a_3 & \cdots & 0 & 0 & 0 \\ \cdot & \cdot & \cdot & \cdots & \cdot & \cdot & \cdot \\ 0 & 0 & 0 & \cdots & b_{n-2} & a_{n-1} & b_{n-1} \\ 0 & 0 & 0 & \cdots & 0 & b_{n-1} & a_n \end{Vmatrix}, \tag{77}$$

where

$$a_i = Q_i \Delta x + \frac{P_{i-1} + P_i}{\Delta x} \qquad (i = 1, \ldots, n) \tag{78}$$

and

$$b_i = -\frac{P_i}{\Delta x} \qquad (i = 1, \ldots, n-1). \tag{79}$$

A symmetric matrix like (77), all of whose elements vanish except those appearing on the principal diagonal and on the two adjoining diagonals, is called a *Jacobi matrix*, and a quadratic form with such a matrix is called a *Jacobi form*. For any Jacobi matrix, there is a recurrence relation between the *descending principal minors*, i.e., between the determinants

$$D_i = \begin{vmatrix} a_1 & b_1 & 0 & \cdots & 0 & 0 & 0 \\ b_1 & a_2 & b_2 & \cdots & 0 & 0 & 0 \\ 0 & b_2 & a_3 & \cdots & 0 & 0 & 0 \\ \cdot & \cdot & \cdot & \cdots & \cdot & \cdot & \cdot \\ 0 & 0 & 0 & \cdots & b_{i-2} & a_{i-1} & b_{i-1} \\ 0 & 0 & 0 & \cdots & 0 & b_{i-1} & a_i \end{vmatrix}, \tag{80}$$

where $i = 1, \ldots, n$. In fact, expanding D_i with respect to the elements of the last row, we obtain the recursion relation

$$D_i = a_i D_{i-1} - b_{i-1}^2 D_{i-2}, \tag{81}$$

which allows us to determine the minors $D_3, \ldots, D_n$ in terms of the first two minors D_1 and D_2. Moreover, if we set $D_0 = 1$, $D_{-1} = 0$, then (81) is valid for all $i = 1, \ldots, n$, and uniquely determines $D_1, \ldots, D_n$.

According to a familiar result, sometimes called the *Sylvester criterion*, a quadratic form

$$\sum_{i, k=1}^{n} a_{ik} \xi_i \xi_k \qquad (a_{ki} = a_{ik})$$

is positive definite if and only if the descending principal minors

$$a_{11}, \quad \begin{vmatrix} a_{11} & a_{12} \\ a_{21} & a_{22} \end{vmatrix}, \quad \begin{vmatrix} a_{11} & a_{12} & a_{13} \\ a_{21} & a_{22} & a_{23} \\ a_{31} & a_{32} & a_{33} \end{vmatrix}, \quad \ldots, \quad \det \| a_{ik} \|$$

of the matrix $\| a_{ik} \|$ are all positive.[26] Applied to the present problem, this criterion states that the Jacobi form (76), with matrix (77), is positive definite if and only if all the quantities defined by (81) are positive, where $i = 1, \ldots, n$ and $D_0 = 1$, $D_{-1} = 0$.

We now use this result to obtain a criterion for the quadratic functional (74) to be positive definite. Thus, we examine what happens to the recurrence relation (81) as $n \to \infty$. Substituting for the coefficients a_i and b_i from (78) and (79), we can write (81) in the form

$$D_i = \left(Q_i \Delta x + \frac{P_{i-1} + P_i}{\Delta x} \right) D_{i-1} - \frac{P_{i-1}^2}{(\Delta x)^2} D_{i-2} \qquad (i = 1, \ldots, n). \tag{82}$$

It is obviously impossible to pass directly to the limit $n \to \infty$ (i.e., $\Delta x \to 0$) in (82), since then the coefficients of D_{i-1} and D_{i-2} become infinite. To avoid this difficulty, we make the "change of variables"[27]

$$D_i = \frac{P_1 \cdots P_i Z_{i+1}}{(\Delta x)^{i+1}} \qquad (i = 1, \ldots, n),$$

$$D_0 = \frac{Z_1}{\Delta x} = 1, \tag{83}$$

$$D_{-1} = Z_0 = 0.$$

---

[26] See e.g., G. E. Shilov, *op. cit.*, Theorem 27, p. 131.

[27] Substituting the expressions (78) and (79) into (80), we find by direct calculation that D_i is of order $(\Delta x)^{-i}$, and hence that Z_i is of order Δx.

In terms of the variables Z_i, the recurrence relation (82) becomes

$$\frac{P_1 \cdots P_i Z_{i+1}}{(\Delta x)^{i+1}} = \left(Q_i \Delta x + \frac{P_{i-1} + P_i}{\Delta x} \right) \frac{P_1 \cdots P_{i-1} Z_i}{(\Delta x)^i}$$
$$- \frac{P_{i-1}^2}{(\Delta x)^2} \frac{P_1 \cdots P_{i-2} Z_{i-1}}{(\Delta x)^{i-1}},$$

i.e.,

$$Q_i Z_i (\Delta x)^2 + P_{i-1} Z_i + P_i Z_i - P_i Z_{i+1} - P_{i-1} Z_{i-1} = 0$$

or

$$Q_i Z_i - \frac{1}{\Delta x} \left(P_i \frac{Z_{i+1} - Z_i}{\Delta x} - P_{i-1} \frac{Z_i - Z_{i-1}}{\Delta x} \right) = 0 \quad (i = 1, \ldots, n). \quad (84)$$

Passing to the limit $\Delta x \to 0$ in (84), we obtain the differential equation

$$- \frac{d}{dx} (PZ') + QZ = 0, \tag{85}$$

which is just the Jacobi equation!

The condition that the quantities D_i satisfying the relation (82) be positive is equivalent to the condition that the quantities Z_i satisfying the difference equation (84) be positive, since the factor

$$\frac{P_1 \cdots P_i}{(\Delta x)^{i+1}}$$

is always positive [because of the condition $P(x) > 0$]. Thus, we have proved that *the quadratic form* (76) *is positive definite if and only if all but the first of the* $n + 2$ *quantities* $Z_0, Z_1, \ldots, Z_{n+1}$ *satisfying the difference equation* (84) *are positive.*[28]

If we consider the polygonal line Π_n with vertices

$$(a, Z_0), \quad (x_1, Z_1), \quad \cdots, \quad (b, Z_{n+1})$$

recall that $a = x_0$, $b = x_{n+1}$), the condition that $Z_0 = 0$ and $Z_i > 0$ for $i = 1, \ldots, n + 1$ means that Π_n does not intersect the interval $[a, b]$ except at the end point a. As $\Delta x \to 0$, the difference equation (84) goes into the Jacobi differential equation (85), and the polygonal line Π_n goes into a nontrivial solution of (85) which satisfies the initial condition

$$Z(a) = Z_0 = 0, \qquad Z'(a) = \lim_{\Delta x \to 0} \frac{Z_1 - Z_0}{\Delta x} = \lim_{\Delta x \to 0} \frac{\Delta x}{\Delta x} = 1$$

and does not vanish for $a < x \leqslant b$. In other words, as $n \to \infty$, the Jacobi form (76) goes into the quadratic functional (74), and the condition that (76)

---

[28] Note that $Z_0 = 0$, $Z_1 = \Delta x > 0$, according to (83). Note also that these two equations, together with the n equations (84), form a system of $n + 2$ independent linear equations in $n + 2$ unknowns, and that such a system always has a unique solution.

be positive definite goes into precisely the condition for (74) to be positive definite given in Theorem 3 of Sec. 26, i.e., the condition that $[a, b]$ contain no points conjugate to a. The legitimacy of this passage to the limit can be made completely rigorous, but we omit the details.

PROBLEMS

1. Calculate the second variation of each of the following functionals:

a) $J[y] = \int_a^b F(x, y)\, dx$;

b) $J[y] = \int_a^b F(x, y, y', \ldots, y^{(n)})\, dx$;

c) $J[u] = \iint_R F(x, y, u, u_x, u_y)\, dx\, dy$.

2. Show that the second variation of a linear functional is zero. State and prove a converse result.

3. Prove that a quadratic functional is twice differentiable, and find its first and second variations.

4. Calculate the second variation of the functional

$$e^{J[y]},$$

where $J[y]$ is a twice differentiable functional.

Ans. $\delta^2 e^{J[y]} = [(\delta J)^2 + \delta^2 J] e^{J[y]}$.

5. Give an example showing that in Theorem 2 of Sec. 24, we cannot replace the condition that $\delta^2 J[h]$ be strongly positive by the condition that $\delta^2 J[h] > 0$.

6. Derive the analog of Legendre's necessary condition for functionals of the form

$$J[u] = \iint_R F(x, y, u, u_x, u_y)\, dx\, dy,$$

where u vanishes on the boundary of R.

Ans. The matrix

$$\begin{Vmatrix} F_{u_x u_x} & F_{u_x u_y} \\ F_{u_y u_x} & F_{u_y u_y} \end{Vmatrix}$$

should be nonnegative definite (cf. p. 119).

7. For which values of a and b is the quadratic functional

$$\int_0^a [f'^2(x) - bf^2(x)]\, dx$$

nonnegative for all $f(x)$ such that $f(0) = f(a) = 0$? Deduce an inequality from the answer.

8. Show that the extremals of any functional of the form

$$\int_a^b F(x, y')\, dx$$

have no conjugate points.

9. Prove that if a family of extremals drawn from a given point A has an envelope E, then the point where a given extremal touches E is a conjugate point of A.

10. Investigate the extremals of the functional

$$J[y] = \int_0^a \frac{y}{y'^2}\, dx, \qquad y(0) = 1, y(a) = A,$$

where $0 < a$, $0 < A < 1$. Show that two extremals go through every pair of points $(0, 1)$ and (a, A). Which of these two extremals corresponds to a weak minimum?

Hint. The line $x = 0$ is an envelope of the family of extremals.

11. Prove that the extremal $y = y_1 x / x_1$ corresponds to a weak minimum of both functionals

$$\int_0^{z_1} \frac{dx}{y'}, \qquad \int_0^{z_1} \frac{dx}{y'^2},$$

where $y(0) = 0$, $y(x_1) = y_1$, $x_1 > 0$, $y_1 > 0$.

12. What is the restriction on a if the functional

$$\int_0^a (y'^2 - y^2)\, dx, \qquad y(0) = 0, \quad y(a) = 0$$

is to satisfy the strengthened Jacobi condition? Use two approaches, one based on Jacobi's equation (42) and the other based on Definition 4 (p. 114) of a conjugate point.

13. Is the strengthened Jacobi condition satisfied by the functional

$$J[y] = \int_0^a (y'^2 + y^2 + x^2)\, dx, \qquad y(0) = 0, \quad y(a) = 0$$

for arbitrary a?

Ans. Yes.

14. Let $y = y(x, \alpha, \beta)$ be a general solution of Euler's equation, depending on two parameters α and β. Prove that if the ratio

$$\frac{\partial y / \partial \alpha}{\partial y / \partial \beta}$$

is the same at two points, the points are conjugate.

15. Consider the catenary

$$y = c \cosh\left(\frac{x + b}{c}\right),$$

where b and c are constants. Show that any point on the catenary except the vertex $(-b, c)$ has one and only one conjugate, and show that the tangents to any pair of conjugate points intersect on the x-axis.

6

FIELDS.
SUFFICIENT CONDITIONS
FOR A STRONG EXTREMUM

In our study of sufficient conditions for a weak extremum, we introduced the important concept of a conjugate point. The simplest and most natural way to introduce this concept is based on the use of families of neighboring extremals (see Sec. 27). Then the conjugate of a point M lying on an extremal γ is defined as the limit of the points of intersection of γ with the neighboring extremals drawn from M.

The utility of studying families of extremals rather than individual extremals is particularly apparent when we turn our attention to the problem of finding sufficient conditions for a *strong* extremum. The study of such families of extremals is intimately connected with the important concept of a *field*, which we introduce in the next section. Since the concept of a field is useful in many problems, we first give a general definition of a field, which is not directly related to variational problems.

31. Consistent Boundary Conditions. General Definition of a Field

Consider a system of second-order differential equations

$$y_i'' = f_i(x, y_1, \ldots, y_n, y_1', \ldots, y_n') \qquad (i = 1, \ldots, n), \tag{1}$$

solved explicitly for the second derivatives. In order to single out a definite solution of this system, we have to specify $2n$ conditions, e.g., boundary conditions of the form

$$y_i' = \psi_i(y_1, \ldots, y_n) \qquad (i = 1, \ldots, n) \tag{2}$$

for two values of x, say x_1 and x_2. Boundary conditions of this kind are commonly encountered in variational problems. If we require that the boundary conditions (2) hold only at one point, they determine a solution of the system (1) which depends on n parameters.

We now introduce the following definitions:

DEFINITION 1. *The boundary conditions*

$$y_i' = \psi_i^{(1)}(y_1, \ldots, y_n) \qquad (i = 1, \ldots, n), \tag{3}$$

prescribed for $x = x_1$, and the boundary conditions

$$y_i' = \psi_i^{(2)}(y_1, \ldots, y_n) \qquad (i = 1, \ldots, n), \tag{4}$$

prescribed for $x = x_2$, are said to be (mutually) consistent if every solution of the system (1) satisfying the boundary conditions (3) at $x = x_1$ also satisfies the boundary conditions (4) at $x = x_2$, and conversely.[1]

DEFINITION 2. *Suppose the boundary conditions*

$$y_i' = \psi_i(x, y_1, \ldots, y_n) \qquad (i = 1, \ldots, n) \tag{5}$$

(where the ψ_i are continuously differentiable functions) are prescribed for every x in the interval $[a, b]$, and suppose they are consistent for every pair of points x_1, x_2 in $[a, b]$. Then the family of mutually consistent boundary conditions (5) is called a field (of directions) for the given system (1).

As is clear from (5), boundary conditions prescribed for every value of x define a system of first-order differential equations. The requirement that the boundary conditions be consistent for different values of x means that the solutions of the system (5) must also satisfy the system (1), i.e., that (1) is implied by (5).

Because of the existence and uniqueness theorem for systems of differential equations,[2] one and only one integral curve of the system (5) passes through

---

[1] Thus, one might say that the boundary conditions at x_1 can be replaced by the boundary conditions at x_2 which are consistent with those at x_1. In a boundary value problem, the boundary conditions represent the influence of the external medium. But in every concrete problem, we are at liberty to decide what is taken to be the external medium and what is taken to be the system under consideration. For example, in studying a vibrating string, subject to certain boundary conditions at its end points, we can focus our attention on a part of the string, instead of the whole string, regarding the rest of the string as part of the external medium and replacing the effect of the "discarded" part of the string by suitable boundary conditions at the end points of the "retained" part of the string.

[2] See e.g., E. A. Coddington, *op. cit.*, Chap. 6.

each point $(x, y_1, \ldots, y_n)$ of the region R where the functions $\psi_i(x, y_1, \ldots, y_n)$ are defined. According to what has just been said, each of these curves is at the same time a solution of the system (1). Thus, specifying a field (5) of the system (1) in some region R defines an n-parameter family of solutions of (1), such that one and only one curve from the family passes through each point of R. The curves of the family will be called *trajectories* of the field.[3]

The following theorem gives conditions which must be satisfied by the functions $\psi_i(x, y_1, \ldots, y_n)$, $1 \leqslant i \leqslant n$, if the system (5) is to be a field for the system (1):

THEOREM. *The first-order system*

$$y_i' = \psi_i(x, y_1, \ldots, y_n) \qquad (a \leqslant x \leqslant b; 1 \leqslant i \leqslant n) \tag{6}$$

is a field for the second-order system

$$y_i'' = f_i(x, y_1, \ldots, y_n, y_1', \ldots, y_n') \tag{7}$$

if and only if the functions $\psi_i(x, y_1, \ldots, y_n)$ satisfy the following system of partial differential equations, called the Hamilton-Jacobi system[4] for the original system (7):

$$\frac{\partial \psi_i}{\partial x} + \sum_{k=1}^{n} \frac{\partial \psi_i}{\partial y_k} \psi_k = f_i(x, y_1, \ldots, y_n, \psi_1, \ldots, \psi_n). \tag{8}$$

Thus, every solution of the Hamilton-Jacobi system (8) *gives a field for the original system* (7)

Proof. Differentiating (6) with respect to x, we obtain

$$y_i'' = \frac{\partial \psi_i}{\partial x} + \sum_{k=1}^{n} \frac{\partial \psi_i}{\partial y_k} \frac{dy_k}{dx},$$

i.e.,

$$y_i'' = \frac{\partial \psi_i}{\partial x} + \sum_{k=1}^{n} \frac{\partial \psi_i}{\partial y_k} \psi_k.$$

Thus, the system (7) is a consequence of the system (6) if and only if (8) holds.

Example 1. Consider a single linear differential equation

$$y'' = p(x)y. \tag{9}$$

---

[3] A field is usually defined not as a family of boundary conditions which are compatible at every two points, but as a set of integral curves of the system (1) which satisfy the conditions (5) at every point, i.e., as a general solution of the system (5). However, it seems to us that our definition has certain advantages, in particular, when applying the concept of a field to variational problems involving multiple integrals.

[4] For an explanation of the connection between the system (8) and the Hamilton-Jacobi equation defined in Chapter 4, see the remark on p. 143.

The corresponding Hamilton-Jacobi system reduces to a single equation

$$\frac{\partial \psi}{\partial x} + \frac{\partial \psi}{\partial y} \psi = p(x)y,$$

i.e.,

$$\frac{\partial \psi}{\partial x} + \frac{1}{2} \frac{\partial \psi^2}{\partial y} = p(x)y. \tag{10}$$

The set of solutions of (10) depends on an arbitrary function, and according to the theorem, each of these solutions is a field for equation (9).

The simplest solutions of (10) are those that are linear in y:

$$\psi(x, y) = \alpha(x)y. \tag{11}$$

Substituting (11) into (10), we obtain

$$\alpha'(x)y + \alpha^2(x)y = p(x).$$

Thus, $\alpha(x)$ satisfies the Riccati equation

$$\alpha'(x) + \alpha^2(x) = p(x). \tag{12}$$

Solving (12) and setting

$$y' = \alpha(x)y,$$

we obtain a field (which is linear in y) for the differential equation (9).

Example 2. In the same way, we can find the simplest field for a system of linear differential equations

$$Y'' = P(x)Y, \tag{13}$$

where $Y = (y_1, \ldots, y_n)$ and $P(x) = \|p_{ik}(x)\|$ is a matrix. The system of Hamilton-Jacobi equations corresponding to (13) is

$$\frac{\partial \psi_i}{\partial x} + \sum_{k=1}^{n} \frac{\partial \psi_i}{\partial y_k} \psi_k = \sum_{k=1}^{n} p_{ik}(x)y_k \qquad (i = 1, \ldots, n). \tag{14}$$

Let us look for a solution of (14) which is linear in Y, i.e.,

$$\psi_i(x, y_1, \ldots, y_n) = \sum_{k=1}^{n} \alpha_{ik}(x)y_k, \tag{15}$$

or in vector notation,

$$\Psi = AY.$$

Substituting (15) into (14), we obtain

$$\sum_{k=1}^{n} \alpha'_{ik}(x)y_k + \sum_{k=1}^{n} \alpha_{ik}(x) \sum_{j=1}^{n} \alpha_{kj}(x)y_j = \sum_{k=1}^{n} p_{ik}(x)y_k,$$

or in matrix form

$$\left[\frac{d}{dx} A(x) \right] Y + A^2(x)Y = P(x)Y,$$

where $A = \|\alpha_{ik}\|$. Thus, if the matrix $A(x)$ satisfies the equation

$$\frac{d}{dx} A(x) + A^2(x) = P(x),$$

which it is natural to call a *matrix Riccati equation* (cf. p. 120), the functions (15) define a field for the system (13), and this field is linear in y.

It is worth noting, although this observation will not be needed later, that the concept of a field is intimately related to the solution of boundary value problems for systems of second-order differential equations by the so-called "sweep method." We illustrate this method by considering the very simple case where the system consists of a single linear differential equation

$$y''(x) = p(x)y(x) + f(x), \tag{16}$$

with the boundary conditions

$$y'(a) = c_0 y(a) + d_0, \tag{17}$$

$$y'(b) = c_1 y(b) + d_1. \tag{18}$$

We begin by constructing the first-order differential equation

$$y'(x) = \alpha(x)y(x) + \beta(x) \tag{19}$$

and requiring that all its solutions satisfy the boundary condition (17) and the original equation (16). Obviously, to meet the first requirement, we must set

$$\alpha(a) = c_0, \qquad \beta(a) = d_0. \tag{20}$$

To meet the second requirement, we differentiate (19), obtaining

$$y''(x) = \alpha'(x)y(x) + \alpha(x)y'(x) + \beta'(x).$$

Substituting (19) for $y'(x)$ in the right-hand side, we find that

$$y''(x) = [\alpha'(x) + \alpha^2(x)]y(x) + \beta'(x) + \alpha(x)\beta(x),$$

from which it is clear that (19) implies (16) if

$$\begin{aligned} \alpha'(x) + \alpha^2(x) &= p(x), \\ \beta'(x) + \alpha(x)\beta(x) &= f(x). \end{aligned} \tag{21}$$

Now let $\alpha(x)$ and $\beta(x)$ be a solution of the system (21), satisfying the initial conditions (20). Once we have found $\alpha(x)$ and $\beta(x)$, we can write a "boundary condition"

$$y'(x_0) = \alpha(x_0)y(x_0) + \beta(x_0)$$

for every point x_0 in $[a, b]$. This process of shifting the boundary condition originally prescribed for $x = a$ over to every other point in the interval [a, b] is called the "forward sweep." In particular, setting $x = b$, we obtain the equation

$$y'(b) = \alpha(b)y(b) + \beta(b),$$

which, together with the boundary condition (18), forms a system determining $y(b)$ and $y'(b)$. If these values are uniquely determined, our original boundary value problem has a unique solution, i.e., the solution of equation (19) which for $x = b$ takes the value $y(b)$ just found. This second stage in the solution of the boundary value problem is called the "backward sweep." These considerations apply to the case of a single equation, but a similar method can be used to deal with systems of second-order differential equations.

The use of the sweep method to solve the boundary value problem consisting of the differential equation (16) and the boundary conditions (17) and (18) has decided advantages over the more traditional method. [In the latter method, we first find a general solution of equation (16) and then choose the values of the arbitrary constants appearing in this solution in such a way that the boundary conditions (17) and (18) are satisfied.] These advantages are particularly marked in cases where one must resort to some kind of approximate numerical method in order to solve the problem.[5]

The connection between the sweep method and the concept (introduced earlier) of the field of a system of second-order differential equations is now entirely clear. In fact, in the simple case just considered, the forward sweep is nothing but the construction of a field linear in y for equation (16). Moreover, (21) is just the system of ordinary differential equations to which the Hamilton-Jacobi system reduces in the case where we are looking for a field linear in y of a single second-order differential equation.[6]

We might have constructed a field starting from the right-hand end point of the interval [a, b], rather than from the left-hand end point. Thus, our boundary value problem actually involves two fields for equation (16), one of which is determined by shifting the boundary condition (17) from a to b, and the other by shifting the boundary condition (18) from b to a. The solution of the boundary value problem consisting of the differential equation (16) and the boundary conditions (17) and (18) is a curve which is a common trajectory of these two fields. Thus, in the sweep method, we construct one field (the forward sweep) and then choose one of its trajectories which is simultaneously a trajectory of a second field (the backward sweep).

---

[5] I. S. Berezin and N. P. Zhidkov, Методы Вычислений, Том II (*Computational Methods, Vol. II*), Gos. Izd. Fiz.-Mat. Lit., Moscow (1959), Chap. 9, Sec. 9.

[6] In Example 1, we considered the even simpler homogeneous differential equation $y'' = p(x)y$, and correspondingly, we looked for a field of the homogeneous form $y' = \alpha(x)y$. This led to the Riccati equation (12) for the function $\alpha(x)$, identical with the first of the equations (21).

32. The Field of a Functional

32.1. We now apply the considerations of the preceding section to variational problems. The Euler equations

$$F_{y_i} - \frac{d}{dx} F_{y_i'} = 0 \qquad (i = 1, \ldots, n),$$

corresponding to the functional

$$\int_a^b F(x, y_1, \ldots, y_n, y_1', \ldots, y_n') \, dx, \tag{22}$$

form a system of n second-order differential equations. In order to single out a definite solution of this system, we have to specify $2n$ supplementary conditions, which are usually given in the form of boundary conditions, i.e., relations connecting the values of y_i and y_i' at the end points of the interval $[a, b]$ (there are n such relations at each end point). In many cases, of course, the boundary conditions are determined by the very functional under consideration. For example, consider the variable end point problem for the functional

$$\int_a^b F(x, y_1, \ldots, y_n, y_1', \ldots, y_n') \, dx + g^{(1)}(a, y_1, \ldots, y_n) + g^{(2)}(b, y_1, \ldots, y_n),$$
$$\tag{23}$$

differing from (22) by two functions $g^{(1)}$ and $g^{(2)}$ of the coordinates of the end points of the path along which the functional is considered. Calculating the variation of the functional (23), we obtain

$$\int_a^b \sum_{i=1}^n \left(F_{y_i} - \frac{d}{dx} F_{y_i'} \right) h_i \, dx + \sum_{i=1}^n F_{y_i'} h_i \Big|_{x=a}^{x=b}$$
$$+ \sum_{i=1}^n g_{y_i}^{(1)} h_i(a) + \sum_{i=1}^n g_{y_i}^{(2)} h_i(b). \tag{24}$$

Setting (24) equal to zero, and assuming that the curve $y_i = y_i(x)$, $1 \leqslant i \leqslant n$, is an extremal, we find that

$$\sum_{i=1}^n F_{y_i'} h_i \Big|_{x=a}^{x=b} + \sum_{i=1}^n g_{y_i}^{(1)} h_i(a) + \sum_{i=1}^n g_{y_i}^{(2)} h_i(b) = 0. \tag{25}$$

Since $h_i(a)$ and $h_i(b)$ are arbitrary, (25) implies that

$$(F_{y_i'} - g_{y_i}^{(1)})|_{x=a} = 0 \qquad (i = 1, \ldots, n) \tag{26}$$

and

$$(F_{y_i'} - g_{y_i}^{(2)})|_{x=b} = 0 \qquad (i = 1, \ldots, n). \tag{27}$$

If $g^{(1)} = g^{(2)} \equiv 0$, (25) implies

$$F_{y_i'}|_{x=a} = F_{y_i'}|_{x=b} = 0,$$

i.e., the natural boundary conditions for a variable end point problem like the one considered in Sec. 6 [cf. Chap. 1, formula (29)].[7]

Next, we examine in more detail the boundary conditions corresponding to one end point, say $x = a$. For simplicity, we write g instead of $g^{(1)}$, and adopt the vector notation

$$y = (y_1, \ldots, y_n), \qquad y' = (y'_1, \ldots, y'_n),$$

etc., in arguments of functions (cf. Sec. 29). As usual, we introduce the "momenta" (see footnote 15, p. 86)

$$p_i(x, y, y') = F_{y_i}(x, y, y') \qquad (i = 1, \ldots, n), \tag{28}$$

and then write the boundary conditions (26) in the form

$$p_i(x, y, y')|_{x=a} = g_{y_i}(x, y)|_{x=a} \qquad (i = 1, \ldots, n). \tag{29}$$

The relations (28) determine $y'_1(a), \ldots, y'_n(a)$ as functions of $y_1(a), \ldots, y_n(a)$:[8]

$$y'_i(a) = \psi_i(y)|_{x=a} \qquad (i = 1, \ldots, n). \tag{30}$$

Boundary conditions that can be derived in this way merit a special name:

DEFINITION 1. *Given a functional*

$$\int_a^b F(x, y, y')\, dx,$$

with momenta (28), *the boundary conditions* (30), *prescribed for $x = a$, are said to be self-adjoint if there exists a function $g(x, y)$ such that*

$$p_i[x, y, \psi(y)]|_{x=a} \equiv g_{y_i}(x, y)|_{x=a} \qquad (i = 1, \ldots, n). \tag{31}$$

THEOREM 1. *The boundary conditions* (30) *are self-adjoint if and only if they satisfy the conditions*

$$\frac{\partial p_i[x, y, \psi(y)]}{\partial y_k}\bigg|_{x=a} = \frac{\partial p_k[x, y, \psi(y)]}{\partial y_i}\bigg|_{x=a} \qquad (i, k = 1, \ldots, n), \tag{32}$$

called the self-adjointness conditions.

---

[7] It should also be noted that the boundary conditions corresponding to fixed end points can be regarded as a limiting case of the boundary conditions (26) and (27), although the latter involve the additional functions $g^{(1)}$ and $g^{(2)}$. For example, in the case of the functional

$$\int_a^b F(x, y, y')\, dx - k[y(a) - A]^2,$$

the boundary condition at the left-hand end point is

$$[F_{y'}(x, y, y') - 2k(y - A)]|_{x=a} = 0$$

or

$$y(a) = A + \frac{F_{y'}(x, y, y')}{2k}\bigg|_{x=a}.$$

If we now let $k \to \infty$, we obtain in the limit the boundary condition $y(a) = A$. Similar considerations apply to the case of several functions $y_1, \ldots, y_n$.

[8] The conditions (30) can be thought of as assigning a direction to every point of the hyperplane $x = a$. [Cf. formula (2).]

Proof. If the boundary conditions (30) are self-adjoint, then (31) holds, and hence

$$\frac{\partial p_i[x, y, \psi(y)]}{\partial y_k} = \frac{\partial^2 g(x, y)}{\partial y_i \, \partial y_k} = \frac{\partial p_k[x, y, \psi(y)]}{\partial y_i},$$

which is just (32).   Conversely, if the boundary conditions (30) are such that the functions $p_i[x, y, \psi(y)]$ satisfy (32), then, for $x = a$, the p_i are the partial derivatives with respect to y_i of some function $g(y)$,[9] so that the boundary conditions (30) are self-adjoint in the sense of Definition 1.

Remark. It is immediately clear that for $n = 1$, i.e., in the case of variational problems involving a single unknown function, any boundary condition is self-adjoint, and in fact, the self-adjointness conditions (32) disappear for $n = 1$.

32.2. In the preceding section, we introduced the concept of a field for a system of second-order differential equations. We now define the field of a functional:

DEFINITION 2. *Given a functional*

$$\int_a^b F(x, y, y') \, dx, \tag{33}$$

with the system of Euler equations

$$F_{y_i} - \frac{d}{dx} F_{y_i'} = 0 \qquad (i = 1, \ldots, n), \tag{34}$$

we say that the boundary conditions

$$y_i' = \psi_i^{(1)}(y) \qquad (i = 1, \ldots, n), \tag{35}$$

prescribed for $x = x_1$, and the boundary conditions

$$y_i' = \psi_i^{(2)}(y) \qquad (i = 1, \ldots, n), \tag{36}$$

prescribed for $x = x_2$, are (mutually) consistent with respect to the functional (33) if they are consistent with respect to the system (34), i.e., if every extremal satisfying the boundary conditions (35) at $x = x_1$, also satisfies the boundary conditions (36) at $x = x_2$, and conversely.

DEFINITION 3. *The family of boundary conditions*

$$y_i' = \psi_i(x, y) \qquad (i = 1, \ldots, n), \tag{37}$$

---

[9] See e.g., D. V. Widder, *op. cit.*, Theorem 11, p. 251, and T. M. Apostol, *Advanced Calculus*, Addison-Wesley Publishing Co., Inc., Reading, Mass. (1957), Theorem 10–48, p. 296. (We tacitly assume the required regularity of the functions p_i and of their domain of definition.)

prescribed for every x in the interval $[a, b]$, is said to be a field of the functional (33) *if*

1. *The conditions* (37) *are self-adjoint for every x in $[a, b]$;*
2. *The conditions* (37) *are consistent for every pair of points x_1, x_2 in $[a, b]$.*

In other words, by a field of the functional (33) is meant a field for the corresponding system of Euler equations (34) which satisfies the self-adjointness conditions at every point x. The equations (37) represent a system of first-order differential equations. Its general solution (the family of *trajectories* of the field) is an n-parameter family of extremals such that one and only one extremal passes through each point $(x, y_1, \ldots, y_n)$ of the region where the field is defined.[10]

We now give an effective criterion for a given family of boundary conditions to be the field of a functional:

THEOREM 2.[11] *A necessary and sufficient condition for the family of boundary conditions* (37) *to be a field of the functional* (33) *is that the self-adjointness conditions*

$$\frac{\partial p_i[x, y, \psi(x, y)]}{\partial y_k} = \frac{\partial p_k[x, y, \psi(x, y)]}{\partial y_i} \qquad (38)$$

and the consistency conditions

$$\frac{\partial p_i[x, y, \psi(x, y)]}{\partial x} = -\frac{\partial H[x, y, \psi(x, y)]}{\partial y_i} \qquad (39)$$

be satisfied at every point x in $[a, b]$, where

$$p_i(x, y, y') = F_{y_i'}(x, y, y'), \qquad (40)$$

and H is the Hamiltonian corresponding to the functional (33):

$$H(x, y, y') = -F(x, y, y') + \sum_{i=1}^{n} p_i(x, y, y')y_i'. \qquad (41)$$

Proof. We have already shown in Theorem 1 that the conditions (38) are necessary and sufficient for the boundary conditions

$$y_i' = \psi_i(x, y) \qquad (i = 1, \ldots, n) \qquad (42)$$

---

[10] In the calculus of variations, by a field (of extremals) of a functional is usually meant an n-parameter family of extremals satisfying certain conditions, rather than a family of boundary conditions of the type just described. However, as already remarked (see footnote 3, p. 133), it seems to us that our somewhat different approach to the concept of a field has certain advantages.

[11] This theorem is the analog of the theorem of Sec. 31, and the system of partial differential equations (39) is the analog of the Hamilton-Jacobi system (see p. 133).

to be self-adjoint at every point x in $[a, b]$.   Therefore, it only remains to show that if (38) holds at every point x in $[a, b]$, then the conditions (39) are necessary and sufficient for the boundary conditions (42) to be consistent for $a \leqslant x \leqslant b$.   To prove this, we set

$$y_i' = \psi_i(x, y), \qquad y' = \psi(x, y)$$

in (40) and (41), and substitute the right-hand sides of the resulting equations into (39).   Performing the indicated differentiations and dropping arguments (to keep the notation concise), we obtain

$$F_{y_i'x} + \sum_{k=1}^{n} F_{y_i'y_k} \frac{\partial \psi_k}{\partial x} = F_{y_i} + \sum_{k=1}^{n} F_{y_k'} \frac{\partial \psi_k}{\partial y_i} \\ - \sum_{k=1}^{n} \psi_k \frac{\partial F_{y_k}}{\partial y_i} - \sum_{k=1}^{n} F_{y_k'} \frac{\partial \psi_k}{\partial y_i}. \tag{43}$$

Using the self-adjointness conditions

$$\frac{\partial F_{y_k'}}{\partial y_i} = \frac{\partial F_{y_i'}}{\partial y_k},$$

we can write (43) in the form

$$F_{y_i} = F_{y_i'x} + \sum_{k=1}^{n} F_{y_i'y_k} \frac{\partial \psi_k}{\partial x} + \sum_{k=1}^{n} \psi_k \frac{\partial F_{y_i'}}{\partial y_k} \tag{44}$$

Since

$$\frac{\partial F_{y_i'}}{\partial y_k} = F_{y_i'y_k} + \sum_{j=1}^{n} F_{y_i'y_j'} \frac{\partial \psi_j}{\partial y_k},$$

(44) becomes

$$F_{y_i} = F_{y_i'x} + \sum_{k=1}^{n} F_{y_i'y_k} \psi_k + \sum_{k=1}^{n} F_{y_i'y_k'} \left(\frac{\partial \psi_k}{\partial x} + \sum_{j=1}^{n} \frac{\partial \psi_k}{\partial y_j} \psi_j \right). \tag{45}$$

Along the trajectories of the field, we have

$$\frac{dy_k}{dx} = \psi_k,$$

so that

$$\frac{d^2 y_k}{dx^2} = \frac{\partial \psi_k}{\partial x} + \sum_{j=1}^{n} \frac{\partial \psi_k}{\partial y_j} \psi_j.$$

Therefore, (45) reduces to

$$F_{y_i} = F_{y_i'x} + \sum_{k=1}^{n} F_{y_i'y_k} \frac{dy_k}{dx} + \sum_{k=1}^{n} F_{y_i'y_k'} \frac{d^2 y_k}{dx^2}$$

along the trajectories of the field, or

$$F_{y_i} - \frac{d}{dx} F_{y_i'} = 0, \tag{46}$$

where $1 \leq i \leq n$. This means that the trajectories of the field of directions (42) are extremals, i.e., (42) is a field of the functional

$$\int_a^b F(x, y, y') \, dx, \tag{47}$$

and hence the conditions (39) are sufficient. Since the calculations leading from (39) to (46) are reversible, the conditions (39) are also necessary, and the theorem is proved.

THEOREM 3. *The expression*

$$\frac{\partial p_i(x, y, y')}{\partial y_k} - \frac{\partial p_k(x, y, y')}{\partial y_i} \tag{48}$$

has a constant value along each extremal.

Proof. Using (46), we find that

$$\frac{d}{dx}\left(\frac{\partial p_i}{\partial y_k} - \frac{\partial p_k}{\partial y_i}\right) = \frac{\partial F_{y_i}}{\partial y_k} - \frac{\partial F_{y_k}}{\partial y_i} = 0.$$

COROLLARY. *Suppose the boundary conditions*

$$y_i' = \psi_i(x, y) \qquad (a \leq x \leq b; 1 \leq i \leq n) \tag{49}$$

are consistent, i.e., suppose the solutions of the system (49) *are extremals of the functional* (47). *Then, to prove that the conditions* (49) *define a field of the functional* (47), *it is only necessary to verify that they are self-adjoint at a single (arbitrary) point in* $[a, b]$.

According to Definition 1, the boundary conditions (49) are self-adjoint if there exists a function $g(x, y)$ such that

$$p_i[x, y, \psi(x, y)] = g_{y_i}(x, y) \qquad (i = 1, \ldots, n) \tag{50}$$

for $a \leq x \leq b$. We now ask the following question: What condition has to be imposed on the function $g(x, y)$ in order for the boundary conditions (49), defined by the relations (50), to be not only self-adjoint, but also consistent, at every point of $[a, b]$, i.e., for the boundary conditions (49) to be a field of the functional (47)? The answer is given by

THEOREM 4. *The boundary conditions* (49) *defined by the relations* (50) *are consistent if and only if the function* $g(x, y)$ *satisfies the Hamilton-Jacobi equation*[12]

$$\frac{\partial g}{\partial x} + H\left(x, y_1, \ldots, y_n, \frac{\partial g}{\partial y_1}, \ldots, \frac{\partial g}{\partial y_n}\right) = 0. \tag{51}$$

---

[12] Cf. equation (72), p. 90.

Proof. It follows from (50) that the Hamilton-Jacobi equation (51) can be written in the form

$$\frac{\partial g}{\partial x} = -H(x, y_1, \ldots, y_n, p_1, \ldots, p_n), \qquad (52)$$

where $p_i = p_i[x, y, \psi(x, y)]$. Differentiating (52) with respect to y_i, we obtain

$$\frac{\partial^2 g}{\partial x\, \partial y_i} = -\frac{\partial H[x, y_1, \ldots, y_n, \psi_1(x, y), \ldots, \psi_n(x, y)]}{\partial y_i},$$

i.e.,

$$\frac{\partial p_i}{\partial x} = -\frac{\partial H[x, y_1, \ldots, y_n, \psi_1(x, y), \ldots, \psi_n(x, y)]}{\partial y_i},$$

which is just the set of consistency conditions (39).

Remark. The connection between the Hamilton-Jacobi system introduced in Sec. 31 and the Hamilton-Jacobi equation introduced in Sec. 23 is now apparent. As we saw in Sec. 31, in the case of an arbitrary system of n second-order differential equations, a field is a system of n first-order differential equations of the form (49), where the functions $\psi_i(x, y)$ satisfy the Hamilton-Jacobi system (8). When we deal with the field of a functional, the system (8) turns into the consistency conditions (39), and in this case, we impose the additional requirement that the boundary conditions defining the field be self-adjoint at every point. This means that the field of a functional is not really determined by n functions $\psi_i(x, y)$, but rather by a single function $g(x, y)$ from which the functions $\psi_i(x, y)$ are derived by using the relations (50). In other words, the function $g(x, y)$ is a kind of *potential* for the field of a functional. Since the field of a functional is determined by a single function, instead of by n functions, it is entirely natural that the set of n consistency conditions for such a field should reduce to a single equation, i.e., that the Hamilton-Jacobi system should be replaced by the Hamilton-Jacobi equation.

32.3. Once more, we consider a functional

$$\int_a^b F(x, y, y')\, dx, \qquad (53)$$

whose extremals are curves in the $(n + 1)$-dimensional space of points $(x, y) = (x, y_1, \ldots, y_n)$. Let R be a simply connected region in this space, and let $c = (c_0, c_1, \ldots, c_n)$ be a point lying outside R.

DEFINITION 4. *Let (x, y) be an arbitrary point of R, and suppose that one and only one extremal of the functional (53) leaves c and passes through (x, y), thereby defining a direction*

$$y_i' = \psi_i(x, y) \qquad (i = 1, \ldots, n) \qquad (54)$$

at every point of R. Then the field of directions (54) is called a central field.

THEOREM 5. *Every central field (54) is a field of the functional (53), i.e., satisfies the consistency and self-adjointness conditions.*

Proof. Consider the function

$$g(x, y) = \int_c^{(x, y)} F(x, y, y') \, dx, \tag{55}$$

where the integral is taken along the extremal of (53) joining the point c to the point (x, y). We define a field of directions in R by setting

$$F_{y_i}(x, y, y') \equiv p_i(x, y, y') = g_{y_i}(x, y) \qquad (i = 1, \ldots, n). \tag{56}$$

The theorem will be proved if it can be shown that this field coincides with the original field (54), since then the original field will satisfy the consistency conditions [since its trajectories are extremals] and also the self-adjointness conditions [this follows from Theorem 1 applied to the field defined by (56)]. But (55) is just the function $S(x, y_1, \ldots, y_n)$ of Sec. 23, and hence

$$g_{y_i}(x, y) = p_i(x, y, z),$$

where z denotes the slope of the extremal joining c to (x, y), evaluated at (x, y).[13] This shows that the field of directions (56) actually coincides with the original field (54).

DEFINITION 5. *Given an extremal γ of the functional (53), suppose there exists a simply connected (open) region R containing γ such that*

1. *A field of the functional (53) covers R, i.e., is defined at every point of R;*
2. *One of the trajectories of the field is γ.*

Then we say that γ can be imbedded in a field [of the functional (53)].

THEOREM 6. *Let γ be an extremal of the functional (53), with equation*

$$y = y(x) \qquad (a \leqslant x \leqslant b),$$

in vector form. Moreover, suppose that

$$\det \| F_{y_i'y_k'} \|$$

is nonvanishing in $[a, b]$, and that no points conjugate to $(a, y(a))$ lie on γ. Then γ can be imbedded in a field.

Proof. By hypothesis, the following two conditions are satisfied for sufficiently small $\varepsilon > 0$:

1. The extremal γ can be extended onto the whole interval $[a - \varepsilon, b]$;
2. The interval $[a - \varepsilon, b]$ contains no points conjugate to a (cf. footnote 20, p. 121).

---

[13] See the second of the formulas (70) and footnote 18, p. 90.

Now consider the family of extremals leaving the point $(a - \varepsilon, y(a - \varepsilon))$. Since there are no points conjugate to $a - \varepsilon$ in the interval $[a - \varepsilon, b]$, it follows that for $a \leqslant x \leqslant b$ no two extremals in this family which are sufficiently close to the original extremal γ can intersect. Thus, in some region R containing γ, the extremals sufficiently close to γ define a central field in which γ is imbedded. The proof is now completed by using Theorem 5.

33. Hilbert's Invariant Integral

As before, let R be a simply connected region in the $(n + 1)$-dimensional space of points $(x, y) = (x, y_1, \ldots, y_n)$, and let

$$y_i = \psi_i(x, y) \qquad (i = 1, \ldots, n) \tag{57}$$

define a field of the functional

$$\int_a^b F(x, y, y')\, dx \tag{58}$$

in R. It was proved in the preceding section (see Theorem 2) that the field of directions (57) is a field of the functional (58) if and only if the functions $\psi_i(x, y)$ satisfy the self-adjointness conditions

$$\frac{\partial p_i[x, y, \psi(x, y)]}{\partial y_k} = \frac{\partial p_k[x, y, \psi(x, y)]}{\partial y_i} \tag{59}$$

and the consistency conditions

$$\frac{\partial H[x, y, \psi(x, y)]}{\partial y_i} = -\frac{\partial p_i[x, y, \psi(x, y)]}{\partial x}. \tag{60}$$

Taken together, the conditions (59) and (60) imply that the quantity

$$-H[x, y, \psi(x, y)]\, dx + \sum_{i=1}^n p_i[x, y, \psi(x, y)]\, dy_i$$

is the exact differential of some function (see footnote 9, p. 139)

$$g(x, y) = g(x, y_1, \ldots, y_n).$$

As is familiar from elementary analysis,[14] this function, which is determined to within an additive constant, can be written as a line integral

$$g(x, y) = \int_\Gamma \left(-H\, dx + \sum_{i=1}^n p_i\, dy_i \right), \tag{61}$$

evaluated along the curve Γ going from some fixed point $M_0 = (x_0, y(x_0))$ to the variable point $M = (x, y)$. Since the integrand of (61) is an exact

---

[14] See e.g., D. V. Widder, *op. cit.*, Theorem 12, p. 251.

differential, the choice of the curve Γ does not matter; in fact, the value of the integral depends only on the points M_0, M_1, and not on the curve Γ. The right-hand side of (61) is known as *Hilbert's invariant integral*.

Using the equations (57) defining the field, and explicitly introducing the integrand F of the functional (58), we can write the integral in (61) as

$$\int_{\Gamma} \left(\left\{ F[x, y, \psi(x, y)] - \sum_{i=1}^{n} \psi_i(x, y) F_{y_i'}[x, y, \psi(x, y)] \right\} dx \\ + \sum_{i=1}^{n} F_{y_i'}[x, y, \psi(x, y)] \, dy_i \right). \tag{62}$$

This expression is Hilbert's invariant integral, in the form corresponding to the field defined by the functions $\psi_i(x, y)$. If the curve Γ along which the integral (62) is evaluated is one of the trajectories of the field, then

$$dy_i = \psi_i(x, y) \, dx$$

along Γ, and hence (62) reduces to

$$\int_{\Gamma} F(x, y, y') \, dx$$

evaluated along this trajectory.

Remark. If γ is an extremal which is a trajectory of the field, Hilbert's invariant integral can be used to write the value of the functional for this extremal as an integral evaluated along *any* curve joining the end points of γ. This important fact will be used in the next section.

34. The Weierstrass E-Function. Sufficient Conditions for a Strong Extremum

DEFINITION. *By the Weierstrass E-function of the functional*[15]

$$J[y] = \int_a^b F(x, y, y') \, dx, \qquad y(a) = A, \qquad y(b) = B \tag{63}$$

we mean the following function of $3n + 1$ variables:

$$E(x, y, z, w) = F(x, y, w) - F(x, y, z) - \sum_{i=1}^{n} (w_i - z_i) F_{y_i'}(x, y, z). \tag{64}$$

In other words, $E(x, y, z, w)$ is the difference between the value of the

---

[15] Here $y(a) = A$ means $y_1(a) = A_1, \ldots, y_n(a) = A_n$, and similarly for $y(b) = B$, i.e., we are dealing with the fixed end point problem.

function F (regarded as a function of its last n arguments) at the point w and the first two terms of its Taylor's series expansion about the point z. Thus, $E(x, y, z, w)$ can also be written as the remainder of a Taylor's series:

$$E(x, y, z, w) = \frac{1}{2} \sum_{i, k=1}^{n} (w_i - z_i)(w_k - z_k) F_{y_i' y_k'}[x, y, z + \theta(w - z)]$$

$$(0 < \theta < 1).$$

For $n = 1$, the Weierstrass E-function has a simple geometric interpretation, since if we regard $F(x, y, z)$ as a function of z,

$$F(x, y, w) - F(x, y, z) - (w - z)F_{y'}(x, y, z)$$

is just the vertical distance from the curve Γ representing $F(x, y, z)$ to the tangent to Γ drawn through a fixed point of Γ.

Our goal in this section is to derive sufficient conditions for the functional (63) to have a strong extremum. It will be recalled from Secs. 28 and 29 that the following set of conditions is sufficient for the functional (63) to have a weak minimum[16] for the admissible curve γ:

Condition 1. The curve γ is an extremal;

Condition 2. The matrix $\|F_{y_i' y_k'}\|$ is positive definite along γ;

Condition 3. The interval $[a, b]$ contains no points conjugate to a.

Every strong extremum is simultaneously a weak extremum, but the converse is in general false (see p. 13). Therefore, in looking for sufficient conditions for a strong extremum, it is natural to assume from the outset that the three conditions just listed are satisfied. We then try to supplement them in such a way as to obtain a set of conditions guaranteeing a strong extremum as well as a weak extremum. To find such supplementary conditions, we first recall that Conditions 2 and 3 imply that the given extremal γ can be imbedded in a field

$$y_i' = \psi_i(x, y) \qquad (i = 1, \ldots, n) \tag{65}$$

of the functional (63) [see Theorem 6 of Sec. 32].[17] Let γ have the equations

$$y_i = y_i(x) \qquad (i = 1, \ldots, n),$$

and let γ^* be an arbitrary curve with the same end points as γ, lying in the $(n + 1)$-dimensional region R containing γ and covered by the field (see

---

[16] To be explicit, we consider only conditions for a minimum. To obtain conditions for a maximum, we need only reverse the directions of all inequalities.

[17] The only part of Condition 2 that is used here is the fact that $\det \|F_{y_i' y_k'}\|$ is non-vanishing (in fact, positive) in $[a, b]$.

Definition 5 of Sec. 32). Then, according to equation (62) and the remark at the end of Sec. 33, we have

$$\int_{\gamma} F(x, y, y') = \int_{\gamma*} \left(\left\{ F(x, y, \psi) - \sum_{i=1}^{n} \psi_i F_{y_i'}(x, y, \psi) \right\} dx + \sum_{i=1}^{n} F_{y_i'}(x, y, \psi)\, dy_i \right),$$

(66)

where for simplicity we omit the arguments of the functions ψ and ψ_i. The right-hand side of (66) is just Hilbert's invariant integral, in the form corresponding to the field (65). As usual, we are interested in the increment

$$\Delta J = \int_{\gamma*} F(x, y, y')\, dx - \int_{\gamma} F(x, y, y')\, dx.$$

Using (66), we find that

$$\Delta J = \int_{\gamma*} F(x, y, y')\, dx$$
$$- \int_{\gamma*} \left(\left\{ F(x, y, \psi) - \sum_{i=1}^{n} \psi_i F_i(x, y, \psi) \right\} dx + \sum_{i=1}^{n} F_{y_i'}(x, y, \psi)\, dy_i \right)$$
$$= \int_{\gamma*} \left(F(x, y, y') - F(x, y, \psi) - \sum_{i=1}^{n} (y_i' - \psi_i) F_{y_i'}(x, y, \psi) \right) dx,$$

or in terms of the Weierstrass E-function.[18]

$$\Delta J = \int_{\gamma*} E(x, y, \psi, y')\, dx. \tag{67}$$

We are now in a position to state sufficient conditions for a strong extremum.

THEOREM 1. *Let γ be an extremal, and let*

$$y_i' = \psi_i(x, y) \qquad (i = 1, \dots, n) \tag{68}$$

be a field of the functional

$$J[y] = \int_a^b F(x, y, y')\, dx, \qquad y(a) = A, \qquad y(b) = B. \tag{69}$$

Suppose that at every point $(x, y) = (x, y_1, \dots, y_n)$ of some (open) region containing γ and covered by the field (68),[19] the condition

$$E(x, y, \psi, w) \geqslant 0 \tag{70}$$

is satisfied for every finite vector $w = (w_1, \dots, w_n)$. Then $J[y]$ has a strong minimum for the extremal γ.

---

[18] More explicitly,

$$\Delta J = \int_a^b E(x, y^*, \psi, y^{*\prime})\, dx,$$

where $y_i = y_i^*(x)$ are the equations of the curve γ^*.

[19] By hypothesis, such a region R exists.

Proof. To say that the functional $J[y]$ has a strong minimum for the extremal γ means that ΔJ is nonnegative for any admissible curve γ^* which is sufficiently close to γ in the norm of the space $\mathscr{C}(a, b)$.   But the condition (70) guarantees that the increment ΔJ, given by (67), is non-negative for all such curves.   Note that we do not impose any restrictions at all on the slope of the curve γ^*, i.e., γ^* need not be close to γ in the norm of the space $\mathscr{D}_1(a, b)$.   In fact, γ^* need not even belong to $\mathscr{D}_1(a, b)$.[20]

Remark 1. As already noted, the hypothesis that the extremal γ can be imbedded in a field can be replaced by Conditions 2 and 3.

Remark 2. Since the Weierstrass E-function can be written in the form

$$E(x, y, \psi, w) = \frac{1}{2} \sum_{i, k=1}^{n} (w_i - \psi_i)(w_k - \psi_k) F_{y_i'y_k'} [x, y, \psi + \theta(w - \psi)]$$

$$(0 < \theta < 1)$$

(see p. 147), we can replace (70) by the condition that at every point of some region R containing γ, the matrix $\|F_{y_i'y_k'}(x, y, z)\|$ be nonnegative definite for every finite z.

We conclude this section by indicating the following *necessary* condition for a strong extremum:

THEOREM 2 (*Weierstrass' necessary condition*).   *If the functional*

$$J[y] = \int_a^b F(x, y, y') \, dx, \qquad y(a) = A, \qquad y(b) = B$$

has a strong minimum for the extremal γ, *then*

$$E(x, y, y', w) \geqslant 0 \tag{71}$$

along γ *for every finite* w.

The idea of the proof is the following: If (71) is not satisfied, there exists a point ξ in $[a, b]$ and a vector q such that

$$E[\xi, y(\xi), y'(\xi), q] < 0, \tag{72}$$

where $y = y(x)$ is the equation of the extremal γ.   It can then be shown that a suitable modification of γ leads to an admissible curve γ^* close to γ in the norm of the space $\mathscr{C}(a, b)$ such that

$$\Delta J = \int_{\gamma *} F(x, y, y') \, dx - \int_{\gamma} F(x, y, y') \, dx < 0, \tag{73}$$

which contradicts the hypothesis the $J[y]$ has a strong minimum for γ. However, the construction of γ^* must be carried out carefully, since all we know is that (72) holds for a suitable q (see Probs. 9 and 10).

---

[20] In problems involving strong extrema of the functional (69), we allow broken extremals, i.e., the admissible curves need only be piecewise smooth (and satisfy the boundary conditions).

PROBLEMS

1. Find the curve joining the points $(-1, -1)$ and $(1, 1)$ which minimizes the functional

$$J[y] = \int_{-1}^{1} (x^2 y'^2 + 12y^2) \, dx.$$

What is the nature of the minimum?

Hint. $\Delta J = J[y + h] - J[y] = \int_{-1}^{1} (x^2 h'^2 + 12h^2) \, dx > 0.$

Ans. $J[y]$ has a strong minimum for $y = x^3$.

2. Find the curve joining the points $(1, 3)$ and $(2, 5)$ which minimizes the functional

$$J[y] = \int_{1}^{2} y'(1 + x^2 y') \, dx.$$

What is the nature of the minimum?

Hint. Again calculate ΔJ.

3. Prove that the segment of the x-axis joining $x = 0$ to $x = \pi$ corresponds to a weak minimum but not a strong minimum of the functional

$$J[y] = \int_{0}^{\pi} y^2(1 - y'^2) \, dx, \qquad y(0) = 0, \quad y(\pi) = 0.$$

Hint. Calculate $J[y]$ for

$$y = \frac{1}{\sqrt{n}} \sin nx.$$

4. Prove that the extrema of the functional

$$\int_{a}^{b} n(x, y)\sqrt{1 + y'^2} \, dx$$

are always strong minima if $n(x, y) \geqslant 0$ for all x and y.

5. Investigate the extrema of the following functionals:

a) $J[y] = \int_{-1}^{2} y'(1 + x^2 y') \, dx, \qquad y(-1) = 1, \quad y(2) = 1;$

b) $J[y] = \int_{0}^{\pi/4} (4y^2 - y'^2 + 8y) \, dx, \qquad y(0) = -1, \quad y(\pi/4) = 0;$

c) $J[y] = \int_{1}^{2} (x^2 y'^2 + 12y^2) \, dx, \qquad y(1) = 1, \quad y(2) = 8;$

d) $J[y] = \int_{0}^{1} (y'^2 + y^2 + 2ye^{2x}) \, dx, \qquad y(0) = \tfrac{1}{3}, \quad y(1) = \tfrac{1}{3}e^2.$

Ans. b) A strong maximum for $y = \sin 2x - 1$; d) A strong minimum for $y = \tfrac{1}{3}e^{2x}$.

6. Prove that $y = bx/a$ is a weak minimum but not a strong minimum of the functional

$$J[y] = \int_{0}^{a} y'^3 \, dx,$$

where $y(0) = 0, \ y(a) = b, \ a > 0, \ b > 0$.

Hint. Examine the corresponding Weierstrass E-function.

7. Show that the extremals which give weak minima in Chap. 5, Prob. 10 do not give strong minima.

8. Show that the extremal $y \equiv 0$ of the functional

$$J[y] = \int_0^1 (ay'^2 - 4byy'^3 + 2bxy'^4) \, dx,$$

where

$$y(0) = 0, \quad y(1) = 0, \quad a > 0, \quad b > 0,$$

satisfies both the strengthened Legendre condition and Weierstrass' necessary condition. Also verify that $y \equiv 0$ can be imbedded in a field of the functional $J[y]$. Does $y \equiv 0$ correspond to a strong minimum of $J[y]$?

   Hint. Choose

$$y = y_0(x) = \begin{cases} \dfrac{k}{h} x & \text{for} \quad 0 \leqslant x \leqslant h, \\[2mm] k \dfrac{1-x}{1-h} & \text{for} \quad h \leqslant x \leqslant 1. \end{cases}$$

Then, given any $k > 0$ however small, there is an $h > 0$ such that $J[y_0] < 0$.
   Ans. No.

9. Complete the proof of Weierstrass' necessary condition, begun on p. 149.
   Hint. By continuity of the E-function, we can always arrange for the point ξ to be an interior point of $[a, b]$. Choose $h > 0$ such that $\xi - h > a$, and construct the function

$$y = y_h(x) = \begin{cases} y(x) + (x - a)Q & \text{for} \quad a \leqslant x \leqslant \xi - h, \\ (x - \xi)q + y(\xi) & \text{for} \quad \xi - h \leqslant x \leqslant \xi, \\ y(x) & \text{for} \quad \xi \leqslant x \leqslant b, \end{cases}$$

where $y = y(x)$ is the equation of the extremal γ, and Q is the vector determined by the condition

$$y(\xi - h) + (\xi - a - h)Q = -qh + y(\xi).$$

Then let $\Delta(h) = J[y_h] - J[y]$. Prove that $\Delta'(0) = E[\xi, y(\xi), y'(\xi), q] < 0$, which, together with $\Delta(0) = 0$, implies that $J[y_h] - J[y] < 0$ for small enough h.

10. Give another proof of Weierstrass' necessary condition, based on the direct use of Hilbert's invariant integral.

   Hint. Let M_1 be the point $(\xi, y(\xi))$. From a point M_0 on γ sufficiently close to M, construct a central field of the functional. Let R be the region covered by this field, and let $\Phi(M)$ be the value of Hilbert's invariant integral evaluated along any curve in R joining M_0 to the variable point M in R. Draw two surfaces σ_2 and σ_1 of the one-parameter family $\Phi(M) = \text{const}$, the first intersecting γ in a point M_2 lying between M_0 and M_1, the second intersecting γ in the point M_1. Moreover, from M_1 draw the straight line with direction q, and let this line intersect σ_2 in a point M_3. Finally, let γ^* be obtained from γ by replacing the part of γ from M_0 to M_1 by the curve $M_0M_3M_1$, where M_0M_3 is the extremal from M_0 to M_3 and M_3M_1 is the straight line segment from M_3 to M_1. Again using Hilbert's invariant integral, prove that γ^* satisfies the inequality (72).

7

VARIATIONAL PROBLEMS
INVOLVING
MULTIPLE INTEGRALS

In this chapter, we discuss a variety of topics pertaining to functionals which depend on functions of two or more variables. Such functionals arise, for example, in mechanical problems involving systems with infinitely many degrees of freedom (strings, membranes, etc.). In our treatment of systems consisting of a finite number of particles (see Chapter 4), we derived the principle of least action and a general method for obtaining conservation laws (Noether's theorem). These methods will now be applied to systems with infinitely many degrees of freedom.

35. Variation of a Functional Defined on a Fixed Region

Consider the functional

$$J[u] = \int \cdots \int_R F(x_1, \ldots, x_n, u, u_{x_1}, \ldots, u_{x_n}) \, dx_1 \cdots dx_n, \tag{1}$$

depending on n independent variables $x_1, \ldots, x_n$, an unknown function u of these variables, and the partial derivatives $u_{x_1}, \ldots, u_{x_n}$ of u. (As usual, it is assumed that the integrand F has continuous first and second derivatives with respect to all its arguments.) We now calculate the variation of (1), assuming that the region R stays fixed, while the function $u(x_1, \ldots, x_n)$ goes into

$$u^*(x_1, \ldots, x_n) = u(x_1, \ldots, x_n) + \varepsilon\psi(x_1, \ldots, x_n) + \cdots, \tag{2}$$

where the dots denote terms of order higher than 1 relative to ε.   By the *variation* δJ of the functional (1), corresponding to the transformation (2), we mean the principal linear part (in ε) of the difference

$$J[u^*] - J[u].$$

For simplicity, we write $u(x), \psi(x)$ instead of $u(x_1, \ldots, x_n), \psi(x_1, \ldots, x_n)$, dx instead of $dx_1 \cdots dx_n$, etc.   Then, using Taylor's theorem, we find that

$$\begin{aligned}
J[u^*] - J[u] &= \int_R \{F[x, u(x) + \varepsilon\psi(x), u_{x_1}(x) + \varepsilon\psi_{x_1}(x), \ldots, u_{x_n}(x) + \varepsilon\psi_{x_n}(x) \\
&\quad - F[x, u(x), u_{x_1}(x), \ldots, u_{x_n}(x)]\} \, dx \\
&= \varepsilon \int_R \left(F_u + \sum_{i=1}^{n} F_{u_{x_i}}\psi_{x_i}\right) dx + \cdots,
\end{aligned}$$

where the dots again denote terms of order higher than 1 relative to ε.   It follows that

$$\delta J = \varepsilon \int_R \left(F_u + \sum_{i=1}^{n} F_{u_{x_i}}\psi_{x_i}\right) dx \tag{3}$$

is the variation of the functional (1).

Next, we try to represent the variation of the functional (1) as an integral of an expression of the form

$$G(x)\psi(x) + \text{div}\,(\cdots),$$

i.e., we try to transform the expression (3) in such a way that the derivatives ψ_{x_i} only appear in a combination of terms which can be written as a divergence.   To achieve this, we replace

$$F_{u_{x_i}}\psi_{x_i}(x)$$

by

$$\frac{\partial}{\partial x_i} \cdot [F_{u_{x_i}}\psi(x)] - \frac{\partial F_{u_{x_i}}}{\partial x_i}\psi(x)$$

in (3), obtaining

$$\delta J = \varepsilon \int_R \left(F_u - \sum_{i=1}^{n} \frac{\partial}{\partial x_i} F_{u_{x_i}}\right)\psi(x)\, dx + \varepsilon \int_R \sum_{i=1}^{n} \frac{\partial}{\partial x_i}[F_{u_{x_i}}\psi(x)]\, dx. \tag{4}$$

This expression for the variation δJ has the important feature that its second term is the integral of a divergence, and hence can be reduced to an integral over the boundary Γ of the region R.   In fact, let $d\sigma$ be the area of a variable element of Γ, regarded as an $(n - 1)$-dimensional surface.   Then the n-dimensional version of Green's theorem states that

$$\int_R \sum_{i=1}^{n} \frac{\partial}{\partial x_i}[F_{u_{x_i}}\psi(x)]\, dx = \int_\Gamma \psi(x)(G, \nu)\, d\sigma, \tag{5}$$

where

$$G = (F_{u_{x_1}}, \ldots, F_{u_{x_n}})$$

is the n-dimensional vector whose components are the derivatives $F_{u_{x_i}}$, $v = (v_1, \ldots, v_n)$ is the unit outward normal to Γ, and (G, v) denotes the scalar product of G and v. Using (5), we can write (4) in the form

$$\delta J = \varepsilon \int_R \left(F_u - \sum_{i=1}^n \frac{\partial}{\partial x_i} F_{u_{x_i}} \right) \psi(x)\, dx + \varepsilon \int_\Gamma \psi(x)(G, v)\, d\sigma, \tag{6}$$

where the integral over R no longer involves the derivatives of $\psi(x)$.

In order for the functional (1) to have an extremum, we must require that $\delta J = 0$ for all admissible $\psi(x)$, in particular, that $\delta J = 0$ for all admissible $\psi(x)$ which vanish on the boundary Γ. For such functions, (6) reduces to

$$\delta J = \int_R \left(F_u - \sum_{i=1}^n \frac{\partial}{\partial x_i} F_{u_{x_i}} \right) \psi(x)\, dx,$$

and then, because of the arbitrariness of $\psi(x)$ inside R, $\delta J = 0$ implies that

$$F_u - \sum_{i=1}^n \frac{\partial}{\partial x_i} F_{u_{x_i}} = 0 \tag{7}$$

for all $x \in R$. This is the *Euler equation* of the functional (1), and is the n-dimensional generalization of formula (24) of Sec. 5.[1]

Remark. In deriving (7), we assumed that the region of integration R appearing in the functional (1) is fixed. Generalization of (7) to the case where the region of integration is variable will be made in Sec. 36.

36. Variational Derivation of the Equations of Motion of Continuous Mechanical Systems

As we saw in Sec. 21, the equations of motion of a mechanical system consisting of n particles can be derived from the *principle of least action*, which states that the actual trajectory of the system in phase space minimizes the *action functional*

$$\int_{t_0}^{t_1} (T - U)\, dt, \tag{8}$$

where T is the kinetic energy and U the potential energy of the system of particles. We now use this principle, together with our basic formula for the first variation, to derive the equations of motion and the appropriate boundary conditions for some simple mechanical systems with infinitely many degrees of freedom, namely, the vibrating string, membrane and plate.

---

[1] As we shall see in the next section, boundary conditions for the equation (7) can be obtained by removing the restriction that $\psi(x) = 0$ on Γ, and then setting $\delta J = 0$ after substitution of (7) into (4) or (6).

36.1. The vibrating string. Consider the transverse motion of a *string* (i.e., a homogeneous flexible cord) of length l and linear mass density ρ. Suppose the ends of the string (at $x = 0$ and $x = l$) are *fastened elastically*, which means that if either end is displaced from its equilibrium position, a restoring force proportional to the displacement appears. This can be achieved, for example, by fastening the ends of the string to two rings which are constrained to move along two parallel rods, while the rings themselves are held in their initial positions by two ideal springs,[2] as shown in Fig. 8. Let the equilibrium position of the string lie along the x-axis, and let $u(x, t)$ denote the displacement of the string at the point x and time t from its equilibrium position. Then, at time t, the kinetic energy of the element of string which initially lies between x_0 and $x_0 + \Delta x$ is clearly

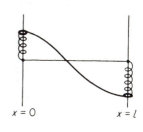

$$\tfrac{1}{2} \rho u_t^2(x_0, t)\,\Delta x. \tag{9}$$

FIGURE 8

Integrating (9) from 0 to l, we find that the kinetic energy of the whole string at time t equals

$$T = \frac{1}{2}\,\rho \int_0^l u_t^2(x, t)\,dx. \tag{10}$$

To find the potential energy of the string, we use the following argument: The potential energy of the string in the position described by the function $u(x, t)$, where t is fixed, is just the work required to move the string from its equilibrium position $u \equiv 0$ into the given position $u(x, t)$. Let τ denote the tension in the spring, and consider the element of string indicated by AB in Figure 9, which initially occupies the position DE along the x-axis, i.e., the interval $[x_0, x_0 + \Delta x]$.[3] To calculate the amount of work needed to move DE to AB, we first move DE to the position AC. This requires no work at all, since the force (the tension in the string) is perpendicular to the displacement.[4] Next, we stretch the string from the position AC to the position AC', where the length of AC' equals the length of AB. This obviously requires an amount of work equal to $\tau\beta$, where β is the length of CC'. Finally, we rotate AC' about the point A into the final position AB. Like the first step, this requires no work at all, since at each stage of the rotation the force is perpendicular to the displacement. Thus, the total amount of work

---

[2] The springs are ideal in the sense that they have zero length when not stretched.

[3] Since we only consider the case of small vibrations, the string can be assumed to have constant length and constant tension. In the present approximation, we can also assume that AB is a straight line segment.

[4] It should be emphasized that since the string is assumed to be absolutely flexible, all the work is expended in stretching the string, and none in bending it.

required to move DE to AB is just the product of τ and the increase in length of the element of string, i.e., the quantity

$$\tau\sqrt{(\Delta x)^2 + (\Delta u)^2} - \tau\Delta x = \frac{1}{2}\tau\left(\frac{\Delta u}{\Delta x}\right)^2 \Delta x + \cdots = \frac{1}{2}\tau u_x^2(x_0, t)\Delta x + \cdots,$$

$$(11)$$

where the dots indicate terms of order higher than those written ($\Delta u/\Delta x \ll 1$ for all t, since the vibrations are small).

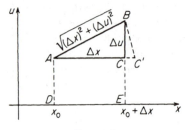

FIGURE 9

Integrating (11) from 0 to l, we find that the potential energy of the whole string is

$$U_1 = \frac{1}{2}\tau\int_0^l u_x^2(x, t)\, dx,\qquad(12)$$

except for the work expended in displacing the elastically fastened ends of the string from their equilibrium positions.   This work equals

$$U_2 = \frac{1}{2}\varkappa_1 u^2(0, t) + \frac{1}{2}\varkappa_2 u^2(l, t),\qquad(13)$$

where $\varkappa_1$ and $\varkappa_2$ are positive constants (the *elastic moduli* of the springs). [In fact, the force f_1 acting on the end point P_1 (see Figure 8) is proportional to the displacement ξ of P_1 from its equilibrium position $x = 0$, $u = 0$, i.e.,

$$|f_1| = \varkappa_1\xi,\qquad(14)$$

where $\varkappa_1 > 0$ is a constant; integration of (14) shows that the work required to move P_1 from $(0, 0)$ to $(0, u(0, t))$, its position at time t, is given by

$$\int_0^{u(0,\, t)} \varkappa_1\xi\, d\xi = \frac{1}{2}\varkappa_1 u^2(0, t),$$

and similarly for the other end point P_2.]   Then, adding (12) and (13), we find that the total potential energy of the string in the position described by the function $u(x, t)$ is

$$U = U_1 + U_2 = \frac{1}{2}\tau\int_0^l u_x^2(x, t)\, dx + \frac{1}{2}\varkappa_1 u^2(0, t) + \frac{1}{2}\varkappa_2 u^2(l, t).\quad(15)$$

Finally, using (10) and (15), we write the action (8) for the vibrating string, obtaining the functional

$$J[u] = \frac{1}{2} \int_{t_0}^{t_1} \int_0^l [\rho u_t^2(x, t) - \tau u_x^2(x, t)] \, dx \, dt \tag{16}$$
$$- \frac{1}{2} \varkappa_1 \int_{t_0}^{t_1} u^2(0, t) \, dt - \frac{1}{2} \varkappa_2 \int_{t_0}^{t_1} u^2(l, t) \, dt.$$

According to the principle of least action, δJ must vanish for the function $u(x, t)$ which describes the actual motion of the string. Thus, we now calculate the variation δJ of the functional (16). Suppose we go from the function $u(x, t)$ to the "varied" function

$$u^*(x, t) = u(x, t) + \varepsilon \psi(x, t) + \cdots$$

Then, using formula (4) and the fact that the variation of a sum equals the sums of the variations of the separate terms, we find that

$$\delta J = \varepsilon \left\{ \int_{t_0}^{t_1} \int_0^l [-\rho u_{tt}(x, t) + \tau u_{xx}(x, t)] \psi(x, t) \, dx \, dt \right.$$
$$\left. - \varkappa_1 \int_{t_0}^{t_1} u(0, t) \, \psi(0, t) \, dt - \varkappa_2 \int_{t_0}^{t_1} u(l, t) \psi(l, t) \, dt \right\} \tag{17}$$
$$+ \varepsilon \int_{t_0}^{t_1} \int_0^l \frac{\partial}{\partial x} [-\tau u_x(x, t) \, \psi(x, t)] \, dx \, dt$$
$$+ \varepsilon \int_{t_0}^{t_1} \int_0^l \frac{\partial}{\partial t} [\rho u_t(x, t) \psi(x, t)] \, dx \, dt.$$

If we assume that the admissible functions $\psi(x, t)$ are such that

$$\psi(x, t_0) = 0, \quad \psi(x, t_1) = 0 \qquad (0 \leqslant x \leqslant l),$$

i.e., that $u(x, t)$ is not varied at the initial and final times, then the last term in (17) vanishes and the next to the last term reduces to

$$\varepsilon \int_{t_0}^{t_1} [\tau u_x(0, t) \psi(0, t) - \tau u_x(l, t) \psi(l, t)] \, dt.$$

It follows that the variation (17) can be written in the form

$$\delta J = \varepsilon \left\{ \int_{t_0}^{t_1} \int_0^l [-\rho u_{tt} + \tau u_{xx}(x, t)] \psi(x, t) \, dx \, dt \right.$$
$$- \int_{t_0}^{t_1} [\varkappa_1 u(0, t) - \tau u_x(0, t)] \psi(0, t) \, dt \tag{18}$$
$$\left. - \int_{t_0}^{t_1} [\varkappa_2 u(l, t) + \tau u_x(l, t)] \psi(l, t) \, dt \right\}.$$

According to the principle of least action, the expression (18) must vanish for the function $u(x, t)$ corresponding to the actual motion of the string. Suppose first that $\psi(x, t)$ vanishes at the end of the string,[5] i.e., that

$$\psi(0, t) = 0, \quad \psi(l, t) = 0 \qquad (t_0 \leqslant t \leqslant t_1). \tag{19}$$

---

[5] If δJ vanishes for all admissible $\psi(x, t)$, it certainly vanishes for all admissible $\psi(x, t)$ satisfying the extra condition (19).

Then (18) reduces to just

$$\delta J = \varepsilon \int_{t_0}^{t_1} \int_0^l \left[-\rho u_{tt}(x, t) + \tau u_{xx}(x, t) \right] \psi(x, t) \, dx \, dt. \tag{20}$$

Setting (20) equal to zero, and using the arbitrariness of the interval $[t_0, t_1]$ and of the function $\psi(x, t)$ for $0 < x < l$, $t_0 < t < t_1$ (cf. the lemma of Sec. 5), we find that

$$u_{tt}(x, t) = a^2 u_{xx}(x, t) \qquad \left(a^2 = \frac{\tau}{\rho} \right) \tag{21}$$

for $0 \leqslant x \leqslant l$ and all t. This result, called the *equation of the vibrating string*, is the Euler equation of the functional

$$\frac{1}{2} \int_{t_0}^{t_1} \int_0^l \left[u_t^2(x, t) - \tau u_x^2(x, t) \right] dx \, dt.$$

Next, we remove the restriction (19). Since $u(x, t)$ must satisfy (21), the first term in (18) vanishes, and we have

$$\delta J = -\varepsilon \Big\{ \int_{t_0}^{t_1} [\varkappa_1 u(0, t) - \tau u_x(0, t)] \psi(0, t) \, dt$$
$$+ \int_{t_0}^{t_1} [\varkappa_2 u(l, t) + \tau u_x(l, t)] \psi(l, t) \, dt \Big\}. \tag{22}$$

This expression must also vanish for the function $u(x, t)$ corresponding to the actual motion of the string. Since $[t_0, t_1]$ is arbitrary and $\psi(0, t)$, $\psi(l, t)$ are arbitrary admissible functions, equating (22) to zero leads to the relations

$$\varkappa_1 u(0, t) - \tau u_x(0, t) = 0 \tag{23}$$

and

$$\varkappa_2 u(l, t) + \tau u_x(l, t) = 0 \tag{24}$$

for all t. Thus, finally, the function $u(x, t)$ which describes the oscillations of the string must satisfy (21) and the boundary conditions

$$\alpha u(0, t) + u_x(0, t) = 0 \qquad \left(\alpha = -\frac{\varkappa_1}{\tau} \right) \tag{25}$$

and

$$\beta u(l, t) + u_x(l, t) = 0 \qquad \left(\beta = \frac{\varkappa_2}{\tau} \right), \tag{26}$$

which connect the displacement from equilibrium and the direction of the tangent at each end of the string.

Next, suppose the ends of the string are free, which means that the springs shown in Fig. 8 are absent and the rings fastening the string to the lines $x = 0$, $x = l$ can move up and down freely. Then $\varkappa_1 = \varkappa_2 = 0$, and the boundary conditions (23), (24) become

$$u_x(0, t) = 0, \qquad u_x(l, t) = 0.$$

Thus, at a free end point, the tangent to the string always preserves the same slope (zero) as it had in the equilibrium position.

The case where the ends of the string are fixed, corresponding to the boundary conditions

$$u(0, t) = 0, \qquad u(l, t) = 0, \tag{27}$$

can be regarded as a limit of the case of elastically fastened ends. In fact, let the stiffness of the springs binding the ends of the string to their initial positions increase without limit, i.e., let $\varkappa_1 \to \infty$, $\varkappa_2 \to \infty$. Then, dividing (23) by $\varkappa_1$ and (24) by $\varkappa_2$, and taking this limit, we obtain the conditions (27).

36.2. Least action vs. stationary action. The principle of least action is widely used not only in mechanics, but also in other branches of physics, e.g., in electrodynamics and field theory. However, as already noted (see Remark 2, p. 85), in a certain sense the principle is not quite true. For example, consider a *simple harmonic oscillator*, i.e., a particle of mass m oscillating about an equilibrium position under the action of an elastic restoring force (cf. Chap. 4, Prob. 2). The equation of motion of the particle is

$$m\ddot{x} + \varkappa x = 0, \tag{28}$$

with solution

$$x = C \sin (\omega t + \theta), \tag{29}$$

where

$$\omega = \sqrt{\frac{\varkappa}{m}},$$

and the values of the constants C, θ are determined from the initial conditions. Moreover, the particle has kinetic energy

$$T = \tfrac{1}{2}m\dot{x}^2$$

and potential energy

$$U = \tfrac{1}{2}\varkappa x^2,$$

so that the action is

$$\frac{1}{2} \int_{t_0}^{t_1} (m\dot{x}^2 - \varkappa x^2) \, dt. \tag{30}$$

Equation (28) is the Euler equation of the functional (30), but in general we cannot assert that its solution (29) actually minimizes (30). In fact, consider the solution

$$x = \frac{1}{\omega} \sin \omega t, \tag{31}$$

which passes through the point $x = 0$, $t = 0$ and satisfies the condition $\dot{x}(0) = 1$. The point $(\pi/\omega, 0)$ is conjugate to the point $(0, 0)$, since every extremal satisfying condition $x(0) = 0$ intersects the extremal (31) at $(\pi/\omega, 0)$ [see p. 114].   Since

$$F_{\dot{t}\dot{t}} = m > 0$$

for the functional (30), the extremal (31) satisfies the sufficient conditions for a minimum (in fact, a strong minimum), *provided that*

$$0 \leqslant t \leqslant t_0 < \frac{\pi}{\omega}.$$

However, if we consider time intervals greater than π/ω, we can no longer guarantee that the extremal (31) minimizes the functional (30).

Next, consider a system of n coupled oscillators, with kinetic energy

$$T = \sum_{i,k=1}^{n} a_{ik} \dot{x}_i \dot{x}_k \tag{32}$$

(a quadratic form in the velocities $\dot{x}_i$) and potential energy

$$U = \sum_{i,k=1}^{n} b_{ik} x_i x_k \tag{33}$$

(a quadratic form in the coordinates x_i).   The quadratic form (32) is positive definite (since it is a kinetic energy); therefore, (32) and (33) can be simultaneously reduced to sums of squares by a suitable linear transformation[6]

$$x_i = \sum_{k=1}^{n} c_{ik} q_k \qquad (i = 1, \ldots, n), \tag{34}$$

i.e., substitution of (34) into (32) and (33) gives

$$T = \sum_{i=1}^{n} \dot{q}_i^2, \qquad U = \sum_{i=1}^{n} \lambda_i q_i^2.$$

Then the equations of motion of the system of oscillators are given by the Euler equations

$$\frac{d}{dt}\left(\frac{\partial T}{\partial \dot{q}_i}\right) + \frac{\partial U}{\partial q_i} = \ddot{q}_i + \lambda_i q_i = 0 \qquad (i = 1, \ldots, n), \tag{35}$$

corresponding to the action functional

$$\int_{t_0}^{t_1} \sum_{i=1}^{n} (\dot{q}_i^2 - \lambda_i q_i^2)\, dt.$$

---

[6] See e.g., G. E. Shilov, *op. cit.*, Secs. 72 and 73.   The coordinates q_i are often called *normal coordinates*, and the corresponding frequencies ω_i are called *natural frequencies*.

Suppose all the λ_i are positive, which means that we are considering oscillations of the system about a position of stable equilibrium. Then the solution of the system (35) has the form

$$q_i = C_i \sin \omega_i(t + \theta_i) \qquad (i = 1, \ldots, n), \qquad (36)$$

where

$$\omega_i = \sqrt{\lambda_i},$$

and the values of the constants C_i, θ_i are determined from the initial conditions. An argument like that made for the simple harmonic oscillator ($n = 1$) shows that a trajectory of the system [i.e., a curve given by (36) in a space of $n + 1$ dimensions] whose projection on the time axis is of length no greater than π/ω, where

$$\omega = \max_{1 \leqslant i \leqslant n} \omega_i,$$

contains no conjugate points and satisfies the sufficient conditions for a minimum. However, just as before, we cannot guarantee that a trajectory whose projection on the time axis is of length greater than π/ω actually minimizes the action.

Finally, consider a vibrating string of length l with fixed ends.[7] As shown above, the function $u(x, t)$ describing the oscillations of the string satisfies the equation

$$u_{tt}(x, t) = a^2 u_{xx}(x, t)$$

and the boundary conditions

$$u(0, t) = 0, \qquad u(l, t) = 0.$$

It follows that[8]

$$u(x, t) = \sum_{k=1}^{\infty} C_k(x) \sin \omega_k(t + \theta_k),$$

where

$$\omega_k = \frac{ka\pi}{l}, \qquad (37)$$

and $C_k(x)$, θ_k are determined from the initial conditions. Thus, in a certain sense, a vibrating string can be regarded as a system of infinitely many coupled oscillators, with natural frequencies (37). However, the numbers (37) have no finite upper bound, and hence the analogy with the case of n coupled oscillators leads us to believe that for a vibrating string, there is no

---

[7] Unlike the analysis of a system of n oscillators, the elementary argument that follows is meant to be heuristic rather than rigorous.

[8] See e.g., G. P. Tolstov, *Fourier Series*, translated by R. A. Silverman, Prentice-Hall, Inc., Englewood Cliffs, N. J. (1962), p. 271.

time interval short enough to guarantee that $u(x, t)$ actually minimizes the action functional. Similar arguments can be carried out for other systems with infinitely many degrees of freedom.

Guided by the above considerations, we shall henceforth replace the principle of *least* action by the principle of *stationary* action. In other words, the actual trajectory of a given mechanical system will not be required to minimize the action but only to cause its first variation to vanish.

36.3. The vibrating membrane. Consider the transverse motion of a *membrane* (i.e., a homogeneous flexible sheet) of surface mass density ρ. Let $u(x, y, t)$ denote the displacement from equilibrium of the point (x, y) of the membrane, at time t. The kinetic energy of the membrane at time t is given by

$$T = \frac{1}{2} \rho \int\!\!\int_R u_t^2(x, y, t) \, dx \, dy, \tag{38}$$

where R is the region of the xy-plane occupied by the membrane at rest. The potential energy of the membrane in the position described by the function $u(x, y, t)$, where t is fixed, is just the work required to move the membrane from its equilibrium position $u \equiv 0$ into the given position $u(x, y, t)$. This work is the sum of the work U_1 expended in deforming the membrane and the work U_2 expended in moving the boundary of the membrane, which we assume to be elastically fastened to its equilibrium position.

To calculate U_1, let τ denote the tension in the membrane, and consider the element ΔA of the membrane initially occupying the region $x_0 \leqslant x \leqslant x_0 + \Delta x$, $y_0 \leqslant y \leqslant y_0 + \Delta y$. Then, just as in the case of the string, the work needed to deform ΔA equals the product of τ and the increase in the area of ΔA under deformation, i.e.,

$$\begin{aligned}
\tau \sqrt{(\Delta x)^2 + (\Delta u)^2} \, \sqrt{(\Delta y)^2 + (\Delta u)^2} &- \tau \, \Delta x \, \Delta y \\
&= \frac{1}{2} \tau \left[\left(\frac{\Delta u}{\Delta x} \right)^2 + \left(\frac{\Delta u}{\Delta y} \right)^2 \right] \Delta x \, \Delta y + \cdots, \\
&= \frac{1}{2} \tau [u_x^2(x_0, y_0, t) + u_y^2(x_0, y_0, t)] \Delta x \, \Delta y + \cdots,
\end{aligned} \tag{39}$$

where the dots indicate terms of order higher than those written. Integrating (39) over R, we find that the work required to deform the whole membrane is

$$U_1 = \frac{1}{2} \tau \int\!\!\int_R [u_x^2(x, y, t) + u_y^2(x, y, t)] \, dx \, dy. \tag{40}$$

To calculate U_2, we generalize the argument used to derive (14). If Γ is the boundary of the region R, and s is arc length measured along Γ from some fixed point on Γ, then

$$U_2 = \frac{1}{2} \int_\Gamma \varkappa(s) u^2(s, t) \, ds, \tag{41}$$

where $u(s, t)$ is the displacement of the membrane from equilibrium at the point s and time t, and $\varkappa(s)$ is the linear density of the elastic modulus of the forces retaining the boundary of the membrane.[9]   Combining (38), (40) and (41), we find that the action functional for the vibrating membrane is

$$
\begin{aligned}
J[u] &= \int_{t_0}^{t_1} (T - U_1 - U_2) \, dt \\
&= \frac{1}{2} \int_{t_0}^{t_1} \int \int_R \{\rho u_t^2(x, y, t) - \tau[u_x^2(x, y, t) + u_y^2(x, y, t)]\} \, dx \, dy \, dt \quad (42) \\
&\quad - \frac{1}{2} \int_{t_0}^{t_1} \int_\Gamma \varkappa(s) \, u^2(s, t) \, ds \, dt.
\end{aligned}
$$

Suppose we go from the function $u(x, y, t)$ to the "varied" function

$$
u^*(x, y, t) = u(x, y, t) + \varepsilon\psi(x, y, t) + \cdots
$$

Then, using formula (4) of Sec. 35 and dropping arguments of functions, we find that the variation δJ of the functional (42) is

$$
\begin{aligned}
\delta J &= \varepsilon \int_{t_0}^{t_1} \int \int_R [-\rho u_{tt} + \tau(u_{xx} + u_{yy})]\psi \, dx \, dy \, dt \\
&\quad - \varepsilon \int_{t_0}^{t_1} \int_\Gamma \varkappa u\psi \, ds \, dt - \varepsilon\tau \int_{t_0}^{t_1} \int \int_R \left[\frac{\partial}{\partial x}(u_x\psi) + \frac{\partial}{\partial y}(u_y\psi)\right] dx \, dy \, dt \\
&\quad + \varepsilon \int_{t_0}^{t_1} \int \int_R \frac{\partial}{\partial t}(u_t\psi) \, dx \, dy \, dt. \quad (43)
\end{aligned}
$$

Just as in the case of the vibrating string, we assume that the function $u(x, y, t)$ is not varied at the initial and final times, i.e., that

$$
\psi(x, y, t_0) = \psi(x, y, t_1) \equiv 0. \quad (44)
$$

Because of (44), the last integral in (43) vanishes.   Moreover, using Green's theorem in two dimensions (see p. 23), we have

$$
\begin{aligned}
\int \int_R \left[\frac{\partial}{\partial x}(u_x\psi) + \frac{\partial}{\partial y}(u_y\psi)\right] dx \, dy &= \int_\Gamma (u_x\psi \, dy - u_y\psi \, dx) \\
&= \int_\Gamma \left[\frac{\partial u}{\partial n} \cos \vartheta \cdot \psi \, ds \sin\left(\frac{\pi}{2} + \vartheta\right) - \frac{\partial u}{\partial n} \sin \vartheta \cdot \psi \, ds \cos\left(\frac{\pi}{2} + \vartheta\right)\right] \\
&= \int_\Gamma \frac{\partial u}{\partial n} \psi \, ds,
\end{aligned}
$$

where $\partial/\partial n$ denotes differentiation with respect to n, the outward normal to Γ, and ϑ is the angle between n and the x-axis.   Thus, we can finally write (43) in the form

$$
\begin{aligned}
\delta J &= \varepsilon \int_{t_0}^{t_1} \int \int_R [-\rho u_{tt} + \tau(u_{xx} + u_{yy})]\psi \, dx \, dy \, dt \\
&\quad - \varepsilon \int_{t_0}^{t_1} \int_\Gamma \left(\varkappa u + \tau \frac{\partial u}{\partial n}\right)\psi \, ds \, dt.
\end{aligned} \quad (45)
$$

---

[9] More precisely, let the parametric equations of Γ be

$$
x = x(s), \qquad y = y(s), \qquad s_0 \leq s \leq s_1.
$$

Then $u(s, t)$ means $u[x(s), y(s), t]$, and "the point s" means the point $(x(s), y(s))$.

We first assume that

$$\psi(s, t) = 0 \qquad (s \in \Gamma), \tag{46}$$

where t is arbitrary, i.e., that u does not vary on the boundary of the membrane. Then (45) reduces to just

$$\delta J = \int_{t_0}^{t_1} \int\int_R [-\rho u_{tt} + \tau(u_{xx} + u_{yy})] \, dx \, dy \, dt. \tag{47}$$

Setting (47) equal to zero, and using the arbitrariness of the interval $[t_0, t_1]$ and of the function $\psi = \psi(x, y, t)$ inside $R \times [t_0, t_1]$, we find that

$$u_{tt}(x, y, t) = a^2[u_{xx}(x, y, t) + u_{yy}(x, y, t)] \qquad \left(a^2 = \frac{\tau}{\rho}\right) \tag{48}$$

for $(x, y) \in R$ and all t, a result known as the *equation of the vibrating membrane*.[10] Equation (48) can also be written as

$$u_{tt}(x, y, t) = a^2 \nabla^2 u(x, y, t),$$

in terms of the *Laplacian (operator)*

$$\nabla^2 = \frac{\partial^2}{\partial x^2} + \frac{\partial^2}{\partial y^2}. \tag{49}$$

Next, we remove the restriction (46). Since $u(x, y, t)$ must satisfy (48), the first term in (45) vanishes, and we are left with

$$\delta J = -\varepsilon \int_{t_0}^{t_1} \int_\Gamma \left[\varkappa(s)u(s, t) + \tau \frac{\partial u(s, t)}{\partial n} \right] \psi(s, t) \, ds \, dt. \tag{50}$$

Then, since $\psi(s, t)$ is an arbitrary admissible function, equating (50) to zero leads to the formula[11]

$$\varkappa(s)u(s, t) + \tau \frac{\partial u(s, t)}{\partial n} = 0 \qquad (s \in \Gamma). \tag{51}$$

This is the boundary condition satisfied by a vibrating membrane when its boundary is elastically fastened to its equilibrium position. In particular, if the boundary of the membrane is free, $\varkappa(s) = 0$ and (51) becomes

$$\frac{\partial u(s, t)}{\partial n} = 0 \qquad (s \in \Gamma), \tag{52}$$

while if the boundary of the membrane is fixed, $\varkappa(s) = \infty$ and (51) becomes

$$u(s, t) = 0 \qquad (s \in \Gamma). \tag{53}$$

---

[10] By $R \times [t_0, t_1]$ is meant the *Cartesian product* of R and $[t_0, t_1]$, i.e., the set of all points (x, y, t) where $(x, y) \in R$ and $t \in [t_0, t_1]$.

[11] The boundary conditions (51), (52) and (53) hold for all t.

36.4. The vibrating plate. Finally, we use the principle of stationary action to derive the equation of motion and the boundary conditions for the transverse vibrations of a *plate* (i.e., a homogeneous two-dimensional elastic body) with surface mass density ρ. As in the case of the vibrating membrane, let $u(x, y, t)$ denote the displacement from equilibrium of the point (x, y) of the plate, at time t. Then the kinetic energy of the plate at time t is given by

$$T = \frac{1}{2} \rho \iint_R u_t^2(x, y, t) \, dx \, dy, \tag{54}$$

where R is the region of the xy-plane occupied by the plate at rest [cf. (38)].

The potential energy of deformation of the plate, which we denote by U_1, depends on how the plate is bent, and hence involves the second derivatives u_{xx}, u_{xy} and u_{yy}. Unlike the case of the membrane, it is assumed that no work is done in stretching the plate, so that U_1 does not involve u_x and u_y. Moreover, we require U_1 to be a quadratic functional in u_{xx}, u_{xy} and u_{yy},[12] which does not depend on the orientation of the coordinate system. Then, since the matrix

$$\begin{Vmatrix} u_{xx} & u_{xy} \\ u_{yx} & u_{yy} \end{Vmatrix}$$

has just two invariants under rotations, i.e., its trace and its determinant,[13] it follows that

$$U_1 = \iint_R [A(u_{xx} + u_{yy})^2 + B(u_{xx}u_{yy} - u_{xy}^2)] \, dx \, dy, \tag{55}$$

where A and B are constants. Equation (55) is usually written in the form

$$U_1 = \frac{1}{2} c \iint_R [(u_{xx}^2 + u_{yy}^2) - 2(1 - \mu)(u_{xx}u_{yy} - u_{xy}^2)] \, dx \, dy, \tag{56}$$

where c is a constant depending on the choice of units, and μ is an absolute constant (*Poisson's ratio*) characterizing the material from which the plate is made. For simplicity, we set $c = 1$.

In addition to the potential energy of deformation U_1, the total potential energy of the plate may also contain a contribution U_2 due to bending moments with density $m(s, t)$, prescribed on the boundary Γ of R, and a contribution U_3 due to external forces acting on R with surface density $f(x, y, t)$ and on Γ with linear density $p(s, t)$. This would give

$$U_2 = \int_\Gamma m(s, t) \frac{\partial u(s, t)}{\partial n} \, ds, \tag{57}$$

---

[12] This guarantees that the equation of motion of the plate is linear.
[13] See e.g., G. E. Shilov, *op. cit.*, p. 106.

where $\partial/\partial n$ denotes differentiation with respect to n, the outward normal to Γ, and[14]

$$U_3 = \iint_R f(x, y, t)u(x, y, t) \, dx \, dy + \int_\Gamma p(s, t) \, ds. \tag{58}$$

Combining (54), (56), (57) and (58), we find that the action functional for the vibrating plate is

$$\begin{aligned}
J[u] &= \int_{t_0}^{t_1} (T - U_1 - U_2 - U_3) \, dt \\
&= \frac{1}{2} \int_{t_0}^{t_1} \iint_R [\rho u_t^2 - (u_{xx} + u_{yy})^2 + 2(1 - \mu)(u_{xx}u_{yy} - u_{xy}^2) - 2fu] \, dx \, dy \, dt \\
&\quad - \int_{t_0}^{t_1} \int_\Gamma \left(pu + m \frac{\partial u}{\partial n} \right) ds \, dt. \tag{59}
\end{aligned}$$

Unlike the corresponding expressions for the vibrating string and the vibrating membrane, (59) contains second derivatives of the unknown function u. The variation of (59) corresponding to the transition from $u(x, y, t)$ to

$$u^*(x, y, t) = u(x, y, t) + \varepsilon\psi(x, y, t) + \cdots$$

turns out to be (see Problems 4 and 5, p. 190)

$$\begin{aligned}
\delta J &= \varepsilon \int_{t_0}^{t_1} \iint_R (-\rho u_{tt} - \nabla^4 u - f)\psi \, dx \, dy \, dt \\
&\quad + \varepsilon \int_{t_0}^{t_1} \int_\Gamma \left[(P - p)\psi + (M - m) \frac{\partial \psi}{\partial n} \right] ds \, dt. \tag{60}
\end{aligned}$$

Here,

$$M = -[\mu \nabla^2 u + (1 - \mu)(u_{xx}x_n^2 + 2u_{xy}x_n y_n + u_{yy}y_n^2)] \tag{61}$$

and

$$P = \frac{\partial}{\partial n} \nabla^2 u + (1 - \mu) \frac{\partial}{\partial s} [u_{xx}x_n x_s + u_{xy}(x_n y_s + x_s y_n) + u_{yy}y_n y_s], \tag{62}$$

where $\partial/\partial n$ denotes differentiation in the direction of the outward normal to Γ, with direction cosines x_n, y_n, and $\partial/\partial s$ denotes differentiation in the direction of the tangent to Γ, with direction cosines x_s, y_s. Moreover,

$$\nabla^4 u = \nabla^2(\nabla^2 u) = \frac{\partial^4 u}{\partial x^4} + 2 \frac{\partial^4 u}{\partial x^2 \, \partial y^2} + \frac{\partial^4 u}{\partial y^4},$$

according to (49).

We first assume that

$$\psi(s, t) = 0, \quad \frac{\partial \psi(s, t)}{\partial n} = 0 \qquad (s \in \Gamma), \tag{63}$$

---

[14] An identical term might also have been included in the expression for the potential energy of the vibrating membrane.

where t is arbitrary, i.e., that u and its normal derivative do not vary on the boundary of the plate. Then (60) reduces to just

$$\delta J = \varepsilon \int_{t_0}^{t_1} \iint_R (-\rho u_{tt} - \nabla^4 u - f)\psi \, dx \, dy \, dt. \tag{64}$$

Setting (64) equal to zero, and using the arbitrariness of the interval $[t_0, t_1]$ and of the function $\psi = \psi(x, y, t)$ inside $R \times [t_0, t_1]$, we obtain the equation for *forced* vibrations of the plate:[15]

$$\rho u_{tt}(x, y, t) + \nabla^4 u(x, y, t) + f(x, y, t) = 0. \tag{65}$$

If we set $f \equiv 0$, so that there are no external forces acting on the plate, (65) reduces to the equation for *free* vibrations of the plate

$$\rho u_{tt}(x, y, t) + \nabla^4 u(x, y, t) = 0.$$

Finally, if we set $u_{tt} \equiv 0$ in (65) and assume that $f = f(x, y)$ is independent of time, we obtain an equation for the equilibrium position of the plate under the action of external forces:

$$\nabla^4 u(x, y) + f(x, y) = 0.$$

This equation could have been obtained directly from the condition for the potential energy of the plate to have a minimum (see Remark 2 below).

Next, we remove the restriction (63). Since $u(x, y, t)$ must satisfy (65), the first term in (60) vanishes, and we are left with

$$\delta J = \varepsilon \int_{t_0}^{t_1} \int_{\Gamma} \left[(P - p)\psi + (M - m) \frac{\partial \psi}{\partial n} \right] ds \, dt. \tag{66}$$

Then, since the functions ψ, $\partial \psi / \partial n$ and the interval $[t_0, t_1]$ are arbitrary, equating (66) to zero leads to the *natural boundary conditions*

$$P(s, t) - p(s, t) = 0, \quad M(s, t) - m(s, t) = 0 \qquad (s \in \Gamma). \tag{67}$$

If the boundary of the plate is *clamped*, the conditions (67) are replaced by the "imposed" boundary conditions

$$u(s, t) = 0, \quad \frac{\partial u(s, t)}{\partial n} = 0 \qquad (s \in \Gamma).$$

If the plate is *supported*, i.e., if the boundary of the plate is held fixed while the tangent plane at the boundary can vary, we obtain the boundary conditions

$$u(s, t) = 0, \quad M(s, t) - m(s, t) = 0 \qquad (s \in \Gamma).$$

---

[15] When domains of arguments are not specified, it is understood that t is arbitrary and $(x, y) \in R$.

Remark 1. It should be noted that the Euler equation (65) does not involve the coefficient μ. This is explained by the fact that the expression

$$u_{xx}u_{yy} - u_{xy}^2 \tag{68}$$

is the divergence of the vector

$$(u_x u_{yy}, \; - u_x u_{xy}),$$

and hence has no effect on (65). However, (68) does have a decisive effect on the boundary conditions, via the functions $M(s, t)$ and $P(s, t)$.

Remark 2. For a mechanical system to be in equilibrium, its kinetic energy T must vanish and its potential energy U must be independent of time. Under these conditions, the principle of stationary action reduces to the assertion that $\delta U = 0$. Thus, the equilibrium position of the system corresponds to a stationary value of U. Moreover, it can be shown that this stationary value must be a minimum if the equilibrium is to be stable and hence physically realizable. In elasticity theory, this *principle of minimum potential energy* is often replaced by *Castigliano's principle*, which states that the equilibrium position of an elastic body corresponds to a minimum of the work of deformation.[16]

37. Variation of a Functional Defined on a Variable Region

37.1. Statement of the problem. In Sec. 35, we derived a formula for the variation of the functional

$$J[u] = \int \cdots \int_R F(x_1, \ldots, x_n, u, u_{x_1}, \ldots, u_{x_n}) \, dx_1 \cdots dx_n, \tag{69}$$

allowing only the function u (and hence its derivatives) to vary, while leaving the independent variables (and hence the region of integration R) unchanged. We now find the variation of the functional (69) in the general case where the independent variables $x_1, \ldots, x_n$ are varied, as well as the function u and its derivatives. For simplicity, we use vector notation, writing $x = (x_1, \ldots, x_n)$, $dx = dx_1 \cdots dx_n$ and

$$\text{grad } u \equiv \nabla u = (u_{x_1}, \ldots, u_{x_n}).$$

With this notation, (69) becomes

$$J[u] = \int_R F(x, u, \nabla u) \, dx. \tag{70}$$

---

[16] For a detailed treatment of Castigliano's principle and a proof of its equivalence to the principle of minimum potential energy, see e.g., R. Courant and D. Hilbert, *Methods of Mathematical Physics, Vol. I*, Interscience, Inc., New York (1953), pp. 268–272.

Now consider the family of transformations[17]

$$x_i^* = \Phi_i(x, u, \nabla u; \varepsilon),$$
$$u^* = \Psi(x, u, \nabla u; \varepsilon), \tag{71}$$

depending on a parameter ε, where the functions Φ_i $(i = 1, \ldots, n)$ and Ψ are differentiable with respect to ε, and the value $\varepsilon = 0$ corresponds to the identity transformation:

$$\Phi_i(x, u, \nabla u; 0) = x_i,$$
$$\Psi(x, u, \nabla u; 0) = u. \tag{72}$$

The transformation (71) carries the surface σ, with the equation

$$u = u(x) \qquad (x \in R),$$

into another surface σ^*. In fact, replacing u, ∇u in (71) by $u(x)$, $\nabla u(x)$, and eliminating x from the resulting $n + 1$ equations, we obtain the equation

$$u^* = u^*(x^*) \qquad (x^* \in R^*)$$

for σ^*, where $x^* = (x_1^*, \ldots, x_n^*)$, and R^* is a new n-dimensional region. Thus, the transformation (71) carries the functional $J[u(x)]$ into

$$J[u^*(x^*)] = \int_{R^*} F(x^*, u^*, \nabla^* u^*) \, dx^*,$$

where

$$\nabla^* u^* = (u_{x_1^*}^*, \ldots, u_{x_n^*}^*).$$

Our goal in this section is to calculate the variation of the functional (70) corresponding to the transformation from $x, u(x)$ to $x^*, u^*(x^*)$, i.e., the principal linear part (relative to ε) of the difference

$$J[u^*(x^*)] - J[u(x)]. \tag{73}$$

37.2. Calculation of δx_i and δu. As in the proof of Noether's theorem for one-dimensional regions (see p. 82), suppose ε is a small quantity. Then, by Taylor's theorem, we have

$$x_i^* = \Phi_i(x, u, \nabla u; \varepsilon) = \Phi_i(x, u, \nabla u; 0) + \varepsilon \left. \frac{\partial \Phi_i(x, u, \nabla u; \varepsilon)}{\partial \varepsilon} \right|_{\varepsilon = 0} + o(\varepsilon),$$

$$u^* = \Psi(x, u, \nabla u; \varepsilon) = \Psi(x, u, \nabla u; 0) + \varepsilon \left. \frac{\partial \Psi(x, u, \nabla u; \varepsilon)}{\partial \varepsilon} \right|_{\varepsilon = 0} + o(\varepsilon),$$

or using (72),

$$x_i^* = x_i + \varepsilon \varphi_i(x, u, \nabla u) + o(\varepsilon),$$
$$u^* = u + \varepsilon \psi(x, u, \nabla u) + o(\varepsilon), \tag{74}$$

---

[17] These formulas, with n independent variables and 1 unknown function, should be contrasted with the formulas (45) of Sec. 20, with n unknown functions and 1 independent variable.

where

$$\varphi_i(x, u, \nabla u) = \frac{\partial \Phi_i(x, u, \nabla u; \varepsilon)}{\partial \varepsilon}\Bigg|_{\varepsilon=0},$$

$$\psi(x, u, \nabla u) = \frac{\partial \Psi(x, u, \nabla u; \varepsilon)}{\partial \varepsilon}\Bigg|_{\varepsilon=0}. \tag{75}$$

For a given surface σ, with equation $u = u(x)$, (74) leads to the increments

$$\Delta x_i = x_i^* - x_i = \varepsilon\varphi_i(x) + o(\varepsilon) \tag{76}$$

and

$$\Delta u = u^*(x^*) - u(x) = \varepsilon\psi(x) + o(\varepsilon), \tag{77}$$

where we explicitly indicate the arguments x and x^* at which the functions u and u^* are evaluated, and $\varphi_i(x)$, $\psi(x)$ denote the functions (75) with u, ∇u replaced by $u(x)$, $\nabla u(x)$. Formula (77) gives an expression for the change in u-coordinate as we go from the point $(x, u(x))$ on the surface σ to its image $(x^*, u^*(x^*))$ under the transformation (74). The variations δx_i and δu corresponding to (74) are defined as the principal linear parts (relative to ε) of the increments (76) and (77), i.e.,

$$\delta x_i = \varepsilon\varphi_i(x), \qquad \delta u = \varepsilon\psi(x). \tag{78}$$

We must also consider the increment

$$\overline{\Delta u} = u^*(x) - u(x),$$

i.e., the change in u-coordinate as we go from the point $(x, u(x))$ to the point $(x, u^*(x))$ on the surface σ^* *with the same x-coordinate*, where σ^* is the image of the surface σ under the transformation (74). Imitating (77) and (78), we introduce a new function $\overline{\psi}(x)$ and a corresponding variation $\overline{\delta u}$:

$$\overline{\Delta u} = u^*(x) - u(x) = \varepsilon\overline{\psi}(x) + o(\varepsilon),$$

$$\overline{\delta u} = \varepsilon\overline{\psi}(x).$$

To find the relation between ψ and $\overline{\psi}$, or equivalently, between δu and $\overline{\delta u}$, we write

$$\begin{aligned}
\Delta u = u^*(x^*) - u(x) &= [u^*(x^*) - u^*(x)] + [u^*(x) - u(x)] \\
&= \sum_{i=1}^{n} \frac{\partial u^*}{\partial x_i}(x_i^* - x_i) + \overline{\delta u} + o(\varepsilon) \\
&= \sum_{i=1}^{n} \frac{\partial u^*}{\partial x_i}\delta x_i + \overline{\delta u} + o(\varepsilon).
\end{aligned} \tag{79}$$

Since $\partial u^*/dx_i$ and $\partial u/\partial x_i$ differ only by a quantity of order ε, (79) becomes

$$\Delta u \sim \sum_{i=1}^{n} \frac{\partial u}{\partial x_i}\delta x_i + \overline{\delta u},$$

where the symbol $\sim$ denotes equality except for terms of order higher than 1 relative to ε. But $\Delta u \sim \delta u$, since δu is the principal part of Δu, and hence

$$\delta u = \overline{\delta u} + \sum_{i=1}^{n} u_{x_i}\delta x_i. \tag{80}$$

Moreover, since

$$\delta u = \varepsilon\psi, \qquad \overline{\delta u} = \varepsilon\overline{\psi}, \qquad \delta x_i = \varepsilon\varphi_i,$$

(80) also implies

$$\psi = \overline{\psi} + \sum_{i=1}^{n} u_{x_i}\varphi_i. \tag{81}$$

Example. Let u be a function of a single independent variable x, and let (71) be the transformation

$$\begin{aligned} x^* &= x\cos\varepsilon - u(x)\sin\varepsilon = x - \varepsilon u(x) + o(\varepsilon), \\ u^*(x^*) &= x\sin\varepsilon + u(x)\cos\varepsilon = \varepsilon x + u(x) + o(\varepsilon), \end{aligned} \tag{82}$$

i.e., a counterclockwise rotation of the xu-plane about the small angle $\alpha = \varepsilon$. As shown in Figure 10, (82) carries the point $(x,u(x))$ on the curve γ with equation $u = u(x)$ into the point $(x^*, u^*(x^*))$ on its image γ^* with equation $u^* = u^*(x^*)$. It follows from (82) that

$$\delta x = -\varepsilon u(x), \qquad \delta u = \varepsilon x \tag{83}$$

and

$$\varphi(x) = -u(x), \qquad \psi(x) = x. \tag{84}$$

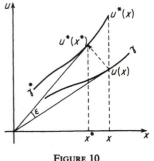

FIGURE 10

In fact, the expressions (83) can be read directly off the figure, as the components of the vector joining the point $(x, u(x))$ to the point $(x^*, u^*(x^*))$. Moreover,

$$u^*(x) = u^*[x^* + \varepsilon u(x)] + o(\varepsilon) = u^*(x^*) + \varepsilon u(x)u^{*\prime}(x^*) + o(\varepsilon),$$

and since $u^{*\prime}(x^*)$ and $u'(x)$ differ only by a quantity of order ε, we have

$$u^*(x) = u^*(x^*) + \varepsilon u(x)u'(x) + o(\varepsilon).$$

On the other hand, according to the second of the formulas (82),

$$u^*(x^*) = \varepsilon x + u(x) + o(\varepsilon).$$

It follows that

$$\overline{\Delta u} = u^*(x) - u(x) = \varepsilon[x + u'(x)u(x)] + o(\varepsilon)$$

and

$$\begin{aligned} \overline{\delta u} &= \varepsilon[x + u(x)u'(x)], \\ \overline{\psi}(x) &= x + u(x)u'(x). \end{aligned} \tag{85}$$

Using (83) and (84), we can write (85) as

$$\begin{aligned} \delta u &= \overline{\delta u} + u'\,\delta x, \\ \psi &= \overline{\psi} + u'\varphi, \end{aligned}$$

in complete agreement with (80) and (81).

37.3. Calculation of δu_{x_i}.   We now derive an expression for the quantity

$$\Delta u_{x_i} = \frac{\partial u^*(x^*)}{\partial x_i^*} - \frac{\partial u(x)}{\partial x_i},$$

or more precisely, its principal part δu_{x_i}, which will be required later when we calculate the increment (73).   First, we note that according to (74),[18]

$$\frac{\partial x_k^*}{\partial x_i} \sim \delta_{ik} + \varepsilon \frac{\partial \varphi_k}{\partial x_i}, \tag{86}$$

where δ_{ik} is the Kronecker delta, equal to 1 if $i = k$ and 0 otherwise.   It follows that

$$\frac{\partial}{\partial x_i} = \sum_{k=1}^{n} \frac{\partial}{\partial x_k^*} \frac{\partial x_k^*}{\partial x_i} \sim \sum_{k=1}^{n} \frac{\partial}{\partial x_k^*} \left(\delta_{ik} + \varepsilon \frac{\partial \varphi_k}{\partial x_i} \right)$$

$$= \frac{\partial}{\partial x_i^*} + \varepsilon \sum_{k=1}^{n} \frac{\partial \varphi_k}{\partial x_i} \frac{\partial}{\partial x_k^*},$$

i.e.,

$$\frac{\partial}{\partial x_i} - \frac{\partial}{\partial x_i^*} \sim \varepsilon \sum_{k=1}^{n} \frac{\partial \varphi_k}{\partial x_i} \frac{\partial}{\partial x_k^*}. \tag{87}$$

Next we write

$$\Delta u_{x_i} = \frac{\partial u^*(x^*)}{\partial x_i^*} - \frac{\partial u(x)}{\partial x_i}$$

$$= \frac{\partial [u^*(x^*) - u(x^*)]}{\partial x_i^*} + \frac{\partial [u(x^*) - u(x)]}{\partial x_i} + \left(\frac{\partial}{\partial x_i^*} - \frac{\partial}{\partial x_i} \right) u(x^*),$$

and analyze each of the three terms in the right-hand side separately.   Using (87) and the fact that

$$u^*(x^*) - u(x^*) \sim \varepsilon \bar{\psi}(x^*),$$

we have

$$\frac{\partial [u^*(x^*) - u(x^*)]}{\partial x_i^*} \sim \frac{\partial [u^*(x^*) - u(x^*)]}{\partial x_i} \sim \varepsilon \frac{\partial \bar{\psi}(x^*)}{\partial x_i} \sim \varepsilon \frac{\partial \bar{\psi}(x)}{\partial x_i}. \tag{88}$$

Moreover, it is easily verified that

$$\frac{\partial [u(x^*) - u(x)]}{\partial x_i} \sim \frac{\partial}{\partial x_i} \sum_{k=1}^{n} \frac{\partial u(x)}{\partial x_k} (x_k^* - x_k) \sim \varepsilon \frac{\partial}{\partial x_i} \sum_{k=1}^{n} \frac{\partial u(x)}{\partial x_k} \varphi_k(x) \tag{89}$$

and

$$\left(\frac{\partial}{\partial x_i^*} - \frac{\partial}{\partial x_i} \right) u(x^*) \sim \left(\frac{\partial}{\partial x_i^*} - \frac{\partial}{\partial x_i} \right) u(x) \sim -\varepsilon \sum_{k=1}^{n} \frac{\partial \varphi_k}{\partial x_i} \frac{\partial u(x)}{\partial x_k^*}. \tag{90}$$

---

[18] In expressions like $\partial \varphi_k / \partial x_i$, u is regarded as a function, i.e., the value of u is not held fixed, as might be inferred from the somewhat ambiguous notation for partial derivatives. Actually, $\partial \varphi_k / \partial x_i$ means

$$\frac{\partial}{\partial x_i} \varphi_k[x, u(x), \nabla u(x)].$$

Adding equations (88), (89) and (90), we obtain

$$\Delta u_{x_i} = \frac{\partial u^*(x^*)}{\partial x_i^*} - \frac{\partial u(x)}{\partial x_i} \sim \varepsilon \left(\frac{\partial \overline{\psi}}{\partial x_i} + \sum_{k=1}^{n} \frac{\partial^2 u}{\partial x_i \, \partial x_k} \varphi_k \right). \tag{91}$$

Finally, recalling that

$$\Delta u_{x_i} \sim \delta u_{x_i}, \qquad \overline{\delta u} = \varepsilon \overline{\psi}, \qquad \delta x_k = \varepsilon \varphi_k,$$

we can write (91) as

$$\delta u_{x_i} = (\overline{\delta u})_{x_i} + \sum_{k=1}^{n} u_{x_i x_k} \, \delta x_k. \tag{92}$$

37.4. Calculation of δJ. We are now in a position to calculate the variation of a functional defined on a variable domain.

THEOREM 1. *The variation of the functional*

$$J[u] = \int_R F(x, u, \nabla u) \, dx \tag{93}$$

corresponding to the transformation[19]

$$\begin{aligned} x_i^* &= \Phi_i(x, u, \nabla u; \varepsilon) \sim x_i + \varepsilon \varphi_i(x, u, \nabla u), \\ u^* &= \Psi(x, u, \nabla u; \varepsilon) \sim u + \varepsilon \psi(x, u, \nabla u) \end{aligned} \tag{94}$$

$(i = 1, \ldots, n)$ *is given by the formula*

$$\delta J = \varepsilon \int_R \left(F_u - \sum_{i=1}^{n} \frac{\partial}{\partial x_i} F_{u_{x_i}} \right) \overline{\psi} \, dx + \varepsilon \int_R \sum_{i=1}^{n} \frac{\partial}{\partial x_i} (F_{u_{x_i}} \overline{\psi} + F \varphi_i) \, dx, \tag{95}$$

where

$$\overline{\psi} = \psi - \sum_{i=1}^{n} u_{x_i} \varphi_i.$$

Proof. Here, δJ means the principal linear part (relative to ε) of the increment

$$\Delta J = J[u^*(x^*)] - J[u(x)], \tag{96}$$

where $u^*(x^*)$ is the image of $u(x)$ under the transformation (94). By definition, (96) equals

$$\begin{aligned} \Delta J &= \int_{R^*} F(x^*, u^*, \nabla^* u^*) \, dx^* - \int_R F(x, u, \nabla u) \, dx \\ &= \int_R \left[F(x^*, u^*, \nabla^* u^*) \frac{\partial(x_1^*, \ldots, x_n^*)}{\partial(x_1, \ldots, x_n)} - F(x, u, \nabla u) \right] dx, \end{aligned} \tag{97}$$

where

$$\frac{\partial(x_1^*, \ldots, x_n^*)}{\partial(x_1, \ldots, x_n)}$$

---

[19] As usual, the symbol $\sim$ denotes equality except for terms of order higher than 1 relative to ε.

is the Jacobian of the transformation from the variables $x_1, \ldots, x_n$ to the variables $x_1^*, \ldots, x_n^*$. According to (86), this Jacobian is

$$\begin{vmatrix} 1 + \varepsilon \dfrac{\partial \varphi_1}{\partial x_1} & \varepsilon \dfrac{\partial \varphi_2}{\partial x_1} & \cdots & \varepsilon \dfrac{\partial \varphi_n}{\partial x_1} \\[2mm] \varepsilon \dfrac{\partial \varphi_1}{\partial x_2} & 1 + \varepsilon \dfrac{\partial \varphi_2}{\partial x_2} & \cdots & \varepsilon \dfrac{\partial \varphi_n}{\partial x_2} \\[2mm] \cdot & \cdot & \cdots & \cdot \\[2mm] \varepsilon \dfrac{\partial \varphi_1}{\partial x_n} & \varepsilon \dfrac{\partial \varphi_2}{\partial x_n} & \cdots & 1 + \varepsilon \dfrac{\partial \varphi_n}{\partial x_n} \end{vmatrix}$$

$$\sim \left(1 + \varepsilon \frac{\partial \varphi_1}{\partial x_1}\right) \cdots \left(1 + \varepsilon \frac{\partial \varphi_n}{\partial x_n}\right) \sim 1 + \varepsilon \sum_{i=1}^{n} \frac{\partial \varphi_i}{\partial x_i},$$

and hence we can write (97) as

$$\Delta J \sim \int_R \left[F(x^*, u^*, \nabla^* u^*) \left(1 + \varepsilon \sum_{i=1}^{n} \frac{\partial \varphi_i}{\partial x_i}\right) - F(x, u, \nabla u) \right] dx. \qquad (98)$$

Using Taylor's theorem to expand the integrand of (98), and retaining only terms of order 1 relative to ε, we find that

$$\delta J = \int_R \left[\sum_{i=1}^{n} F_{x_i} \delta x_i + F_u \delta u + \sum_{i=1}^{n} F_{u_{x_i}} \delta u_{x_i} + \varepsilon F \sum_{i=1}^{n} \frac{\partial \varphi_i}{\partial x_i} \right] dx. \qquad (99)$$

Then, since $\delta x_i = \varepsilon \varphi_i$, substitution of (80) and (92) into (99) gives

$$\delta J = \int_R \left[\sum_{i=1}^{n} F_{x_i} \delta x_i + F_u \overline{\delta u} + F_u \sum_{i=1}^{n} u_{x_i} \delta x_i + \sum_{i=1}^{n} F_{u_{x_i}} (\overline{\delta u})_{x_i} \qquad (100) \right.$$
$$\left. + \sum_{i,k=1}^{n} F_{u_{x_i}} u_{x_i x_k} \delta x_k + F \sum_{i=1}^{n} (\delta x_i)_{x_i} \right] dx.$$

As in the case of a fixed domain R, we try to represent the integrand of (100) as an expression of the form[20]

$$G(x) \overline{\delta u} + \operatorname{div} (\cdots)$$

(cf. p. 153). This can be achieved by noting that

$$\sum_{i=1}^{n} \frac{\partial}{\partial x_i} (F \delta x_i) = \sum_{i=1}^{n} F_{x_i} \delta x_i + \sum_{i=1}^{n} F(\delta x_i)_{x_i}$$
$$+ \sum_{i=1}^{n} F_u u_{x_i} \delta x_i + \sum_{i,k=1}^{n} F_{u_{x_i}} u_{x_i x_k} \delta x_k$$

and

$$\sum_{i=1}^{n} F_{u_{x_i}} (\overline{\delta u})_{x_i} = \sum_{i=1}^{n} \frac{\partial}{\partial x_i} (F_{u_{x_i}} \overline{\delta u}) - \sum_{i=1}^{n} \left(\frac{\partial}{\partial x_i} F_{u_{x_i}} \right) \delta u.$$

---

[20] Then, because of the n-dimensional version of Green's theorem [see formula (5)], the second term of (101) can be transformed into a surface integral.

(The last formula resembles an integration by parts.) Thus, finally, we have

$$\delta J = \int_R \left(F_u - \sum_{i=1}^n \frac{\partial}{\partial x_i} F_{u_{x_i}} \right) \overline{\delta u} \, dx + \int_R \sum_{i=1}^n \frac{\partial}{\partial x_i} (F_{u_{x_i}} \overline{\delta u} + F \, \delta x_i) \, dx, \quad (101)$$

which is the same as formula (95), since $\overline{\delta u} = \varepsilon \overline{\psi}$, $\delta x_k = \varepsilon \varphi_k$. This proves the theorem.

Remark 1. In the special case where the function u and its derivatives are varied, but not the independent variables x_i, we have

$$\varphi_i = 0, \qquad \overline{\psi} = \psi - \sum_{i=1}^n u_{x_i} \varphi_i = \psi,$$

and (95) becomes

$$\delta J = \varepsilon \int_R \left(F_u - \sum_{i=1}^n \frac{\partial}{\partial x_i} F_{u_{x_i}} \right) \psi(x) \, dx + \varepsilon \int_R \sum_{i=1}^n \frac{\partial}{\partial x_i} [F_{u_{x_i}} \psi(x)] \, dx,$$

which is identical with formula (4) of Sec. 35.

Remark 2. The formula for the variation of the functional $J[u]$ is ordinarily used in the case where $u = u(x)$ is an extremal surface of $J[u]$, i.e., satisfies the Euler equation

$$F_u - \sum_{i=1}^n \frac{\partial}{\partial x_i} F_{u_{x_i}} = 0.$$

Then (95) reduces to

$$\delta J = \varepsilon \int_R \sum_{i=1}^n \frac{\partial}{\partial x_i} (F_{u_{x_i}} \overline{\psi} + F \varphi_i) \, dx$$

in the general case, and to

$$\delta J = \varepsilon \int_R \sum_{i=1}^n \frac{\partial}{\partial x_i} (F_{u_{x_i}} \overline{\psi}) \, dx$$

in the case where the independent variables x_i are not varied.

Remark 3. Consider the functional

$$J[u_1, \ldots, u_m] = \int_R F\left(x, u_1, \ldots, u_m, \frac{\partial u_1}{\partial x_1}, \ldots, \frac{\partial u_m}{\partial x_n}\right) dx, \quad (102)$$

involving m unknown functions $u_1, \ldots, u_m$ and their derivatives

$$\frac{\partial u_j}{\partial x_i} \qquad (i = 1, \ldots, n; j = 1, \ldots, m). \quad (103)$$

Introducing the vector $u = (u_1, \ldots, u_m)$ and interpreting ∇u as the tensor with components (103), we can still write (102) in the form

$$J[u] = \int_R F(x, u, \nabla u) \, dx.$$

Then, if (94) is replaced by the transformation

$$x_i^* = \Phi_i(x, u, \nabla u; \varepsilon) \sim x_i + \varepsilon\varphi_i(x, u, \nabla u) \qquad (i = 1, \ldots, n),$$
$$u_j^* = \Psi_j(x, u, \nabla u; \varepsilon) \sim u_j + \varepsilon\psi_j(x, u, \nabla u) \qquad (j = 1, \ldots, m), \qquad (104)$$

the formula (95) generalizes to

$$\delta J = \varepsilon \int_R \sum_{j=1}^{m} \left(F_{u_j} - \sum_{i=1}^{n} \frac{\partial}{\partial x_i} \frac{\partial F}{\partial\left(\frac{\partial u_j}{\partial x_i}\right)} \bar{\psi}_j \right) dx$$
$$+ \varepsilon \int_R \sum_{i=1}^{n} \frac{\partial}{\partial x_i} \left(\sum_{j=1}^{m} \frac{\partial F}{\partial\left(\frac{\partial u_j}{\partial x_i}\right)} \bar{\psi}_j + F\varphi_i \right) dx, \qquad (105)$$

where

$$\bar{\psi}_j = \psi_j - \sum_{i=1}^{n} \frac{\partial u_j}{\partial x_i} \varphi_i \qquad (j = 1, \ldots, m).$$

Remark 4. Let (104) be replaced by the more general transformation

$$x_i^* = \Phi_i(x, u, \nabla u; \varepsilon) \sim x_i + \sum_{k=1}^{r} \varepsilon_k \varphi_i^{(k)}(x, u, \nabla u) \qquad (i = 1, \ldots, n),$$

$$u_j^* = \Psi_j(x, u, \nabla u; \varepsilon) \sim u_j + \sum_{k=1}^{r} \varepsilon_k \psi_j^{(k)}(x, u, \nabla u) \qquad (j = 1, \ldots, m),$$

depending on r parameters $\varepsilon_1, \ldots, \varepsilon_r$, where ε means the vector $(\varepsilon_1, \ldots, \varepsilon_r)$ and the symbol $\sim$ denotes equality except for quantities of order higher than 1 relative to $\varepsilon_1, \ldots, \varepsilon_r$. Then, formula (105) generalizes further to

$$\delta J = \sum_{k=1}^{r} \varepsilon_k \int_R \sum_{j=1}^{m} \left(F_{u_j} - \sum_{i=1}^{n} \frac{\partial}{\partial x_i} \frac{\partial F}{\partial\left(\frac{\partial u_j}{\partial u_i}\right)} \bar{\psi}_j^{(k)} \right) dx$$
$$+ \sum_{k=1}^{r} \varepsilon_k \int_R \sum_{i=1}^{n} \frac{\partial}{\partial x_i} \left(\sum_{j=1}^{m} \frac{\partial F}{\partial\left(\frac{\partial u_j}{\partial x_i}\right)} \bar{\psi}_j^{(k)} + F\varphi_i^{(k)} \right) dx,$$

where

$$\bar{\psi}_j^{(k)} = \psi_j^{(k)} - \sum_{i=1}^{n} \frac{\partial u_j}{\partial x_i} \varphi_i^{(k)} \qquad (k = 1, \ldots, r).$$

37.5. Noether's theorem. Using formula (95) for the variation of a functional, we can deduce an important theorem due to Noether, concerning "invariant variational problems." This theorem has already been proved in Sec. 20 for the case of a single independent variable. Suppose we have a functional

$$J[u] = \int_R F(x, u, \nabla u) \, dx \qquad (106)$$

and a transformation

$$x_i^* = \Phi_i(x, u, \nabla u),$$
$$u^* = \Psi(x, u, \nabla u) \tag{107}$$

$(i = 1, \ldots, n)$ carrying the surface σ with equation $u = u(x)$ into the surface σ^* with equation $u^* = u^*(x^*)$, in the way described on p. 169.

DEFINITION.[21]   *The functional* (106) *is said to be invariant under the transformation* (107) *if $J[\sigma^*] = J[\sigma]$, i.e., if*

$$\int_{R^*} F(x^*, u^*, \nabla^* u^*) \, dx^* = \int_R F(x, u, \nabla u) \, dx.$$

Example.   The functional

$$J[u] = \iint_R \left[\left(\frac{\partial u}{\partial x} \right)^2 + \left(\frac{\partial u}{\partial y} \right)^2 \right] dx \, dy$$

is invariant under the rotation

$$x^* = x \cos \varepsilon - y \sin \varepsilon,$$
$$y^* = x \sin \varepsilon + y \cos \varepsilon, \tag{108}$$
$$u^* = u,$$

where ε is an arbitrary constant.   In fact, since the inverse of the transformation (108) is

$$x = x^* \cos \varepsilon + y^* \sin \varepsilon,$$
$$y = -x^* \sin \varepsilon + y^* \cos \varepsilon,$$
$$u = u^*,$$

it follows that, given a surface σ with equation $u = u(x, y)$, the "transformed" surface σ^* has the equation

$$u^* = u(x^* \cos \varepsilon + y^* \sin \varepsilon, -x^* \sin \varepsilon + y^* \cos \varepsilon) = u^*(x^*, y^*).$$

Consequently, we have

$$J[\sigma^*] = \iint_{R^*} \left[\left(\frac{\partial u^*}{\partial x^*} \right)^2 + \left(\frac{\partial u^*}{\partial y^*} \right)^2 \right] dx^* \, dy^*$$
$$= \iint_{R^*} \left[\left(\frac{\partial u}{\partial x} \cos \varepsilon - \frac{\partial u}{\partial y} \sin \varepsilon \right)^2 + \left(\frac{\partial u}{\partial x} \sin \varepsilon + \frac{\partial u}{\partial y} \cos \varepsilon \right)^2 \right] dx^* \, dy^*$$
$$= \iint_R \left[\left(\frac{\partial u}{\partial x} \right)^2 + \left(\frac{\partial u}{\partial y} \right)^2 \right] \frac{\partial(x^*, y^*)}{\partial(x, y)} \, dx \, dy = \iint_R \left[\left(\frac{\partial u}{\partial x} \right)^2 + \left(\frac{\partial u}{\partial y} \right)^2 \right] dx \, dy.$$

THEOREM 2 (*Noether*).   *If the functional*

$$J[u] = \int_R F(x, u, \nabla u) \, dx \tag{109}$$

---

[21] Cf. the analogous definition on p. 80 and the subsequent examples.

is invariant under the family of transformations

$$x_i^* = \Phi_i(x, u, \nabla u; \varepsilon) \sim x_i + \varepsilon\varphi_i(x, u, \nabla u),$$
$$u^* = \Psi(x, u, \nabla u; \varepsilon) \sim u + \varepsilon\psi(x, u, \nabla u) \qquad (110)$$

$(i = 1, \ldots, n)$ *for an arbitrary region* R, *then*

$$\sum_{i=1}^{n} \frac{\partial}{\partial x_i} (F_{u_{x_i}}\bar{\psi} + F\varphi_i) = 0 \qquad (111)$$

on each extremal surface of $J[u]$, *where*

$$\bar{\psi} = \psi - \sum_{i=1}^{n} u_{x_i}\varphi_i.$$

Proof. According to formula (95),

$$\delta J = \varepsilon \int_R \sum_{i=1}^{n} \frac{\partial}{\partial x_i} (F_{u_{x_i}}\bar{\psi} + F\varphi_i) \, dx,$$

if $u = u(x)$ is an extremal surface. Since $J[u]$ is invariant under (110), $\delta J = 0$, and since R is arbitrary, this implies (111), as asserted.

Remark 1. If we drop the requirement that $u = u(x)$ be an extremal surface of $J[u]$, then, using (95) again, we find that (111) is replaced by

$$\left(F_u - \sum_{i=1}^{n} \frac{\partial}{\partial x_i} F_{u_{x_i}}\right)\bar{\psi} + \sum_{i=1}^{n} \frac{\partial}{\partial x_i} (F_{u_{x_i}}\bar{\psi} + F\varphi_i) = 0.$$

Remark 2. If there are m unknown functions $u_1, \ldots, u_m$, we introduce the vector $u = (u_1, \ldots, u_m)$ and continue to write (109), as in Remark 3, p. 175. Then invariance of $J[u]$ under the family of transformations

$$x_i^* = \Phi_i(x, u, \nabla u; \varepsilon) \sim x_i + \varepsilon\varphi_i(x, u, \nabla u) \qquad (i = 1, \ldots, n),$$
$$u_j^* = \Psi_j(x, u, \nabla u; \varepsilon) \sim u_j + \varepsilon\psi_j(x, u, \nabla u) \qquad (j = 1, \ldots, m)$$

implies that

$$\sum_{i=1}^{n} \frac{\partial}{\partial x_i} \left(\sum_{j=1}^{m} \frac{\partial F}{\partial\left(\dfrac{\partial u_j}{\partial x_i}\right)} \bar{\psi}_j + F\varphi_i \right) = 0, \qquad (112)$$

where

$$\bar{\psi}_j = \psi_j - \sum_{i=1}^{n} \frac{\partial u_j}{\partial x_i} \varphi_i.$$

When $n = 1$, (112) reduces to

$$\frac{d}{dx} \left(\sum_{j=1}^{m} F_{u_j'}\bar{\psi}_j + F\varphi \right) = 0$$

or

$$\sum_{j=1}^{m} F_{u_j'}\bar{\psi}_j + \left(F - \sum_{j=1}^{m} u_j'F_{u_j'}\right)\varphi = \text{const} \qquad (113)$$

along each extremal. This is precisely the version of Noether's theorem proved in Sec. 20. In other words, the left-hand side of (113) is a first integral of the system of Euler equations

$$F_{u_j} - \frac{d}{dx} F_{u'_j} = 0 \qquad (j = 1, \ldots, m).$$

Remark 3. Invariance of the functional (109) under the *r*-parameter family of transformations (see Remark 4, p. 176)

$$x_i^* = \Phi_i(x, u, \nabla u; \varepsilon) \sim x_i + \sum_{k=1}^r \varepsilon_k \varphi_i^{(k)}(x, u, \nabla u) \qquad (i = 1, \ldots, n),$$

$$u_j^* = \Psi_j(x, u, \nabla u; \varepsilon) \sim u_j + \sum_{k=1}^r \varepsilon_k \psi_j^{(k)}(x, u, \nabla u) \qquad (j = 1, \ldots, m)$$

implies the existence of *r* linearly independent relations

$$\sum_{i=1}^n \frac{\partial}{\partial x_i} \left(\sum_{j=1}^m \frac{\partial F}{\partial \left(\dfrac{\partial u_j}{\partial x_i} \right)} \bar{\psi}_j^{(k)} + F \varphi_i^{(k)} \right) = 0 \qquad (k = 1, \ldots, r), \qquad (114)$$

where

$$\bar{\psi}_j^{(k)} = \psi_j^{(k)} - \sum_{i=1}^n \frac{\partial u_j}{\partial x_i} \varphi_i^{(k)}.$$

Remark 4. Suppose the functional $J[u]$ is invariant under a family of transformations depending on *r* arbitrary *functions* instead of *r* arbitrary *parameters.* Then, according to another theorem of Noether (which will not be proved here), there are *r* identities connecting the left-hand sides of the Euler equations corresponding to $J[u]$. For example, consider the simplest variational problem in parametric form, involving a functional

$$J[x, y] = \int_{t_0}^{t_1} \Phi(x, y, \dot{x}, \dot{y}) \, dt, \qquad (115)$$

where Φ is a positive-homogeneous function of degree 1 in $x(t)$ and $y(t)$ (see Sec. 10). Then, as already noted on p. 39, $J[x, y]$ does not change if we introduce a new parameter τ by setting $t = t(\tau)$, where $dt/d\tau > 0$, and in fact, the left-hand sides of the Euler equations

$$\Phi_x - \frac{d}{dt} \Phi_{\dot{x}} = 0, \qquad \Phi_y - \frac{d}{dt} \Phi_{\dot{y}} = 0$$

corresponding to (115) are connected by the identity

$$\dot{x} \left(\Phi_x - \frac{d}{dt} \Phi_{\dot{x}} \right) + \dot{y} \left(\Phi_y - \frac{d}{dt} \Phi_{\dot{y}} \right) = 0.$$

Another interesting example of a family of transformations depending on an arbitrary function, i.e., the *gauge transformations* of electrodynamics, will be given in Sec. 39.

38. Applications to Field Theory

38.1. The principle of stationary action for fields.   In Sec. 36, we discussed the application of the principle of stationary action to vibrating systems with infinitely many degrees of freedom.   These systems were characterized by a function $u(x, t)$ or $u(x, y, t)$ giving the transverse displacement of the system from its equilibrium position.   More generally, consider a physical system (not necessarily mechanical) characterized by one function

$$u(t, x_1, \ldots, x_n) \tag{116}$$

or by a set of functions

$$u_j(t, x_1, \ldots, x_n) \qquad (j = 1, \ldots, m),$$

depending on the time t and the space coordinates $x_1, \ldots, x_n$.[22]   Such a system is called a field [not to be confused with the concept of a field (of directions) treated in Chap. 6], and the functions u_j are called the *field functions*.   As usual, we can simplify the notation by interpreting (116) as a vector function $u = (u_1, \ldots, u_m)$ in the case where $m > 1$.   It is also convenient to write

$$t = x_0, \quad x = (x_0, x_1, \ldots, x_n), \quad dx = dx_0 \, dx_1 \cdots dx_n.$$

Then the field function (116) becomes simply $u(x)$.

In the case of the simple vibrating systems studied in Sec. 36, the equations of motion for the system were derived by first calculating the action functional

$$\int_a^b (T - U) \, dt,$$

where T is the kinetic energy and U the potential energy of the system, and then invoking the principle of stationary action.   Similarly, many other physical fields can be derived from a suitably defined action functional. By analogy with the vibrating string and the vibrating membrane, we write the action in the form[23]

$$J[u, \nabla u] = \int_a^b dx_0 \int \cdots \int_R L(u, \nabla u) \, dx_1 \cdots dx_n = \int_\Omega \mathscr{L}(u, \nabla u) \, dx, \tag{117}$$

---

[22] We deliberately write the argument t first, since it will soon be denoted by x_0. In physical problems, n can only take the values 1, 2 or 3.   However, the choice of m is not restricted, corresponding to the possibility of scalar fields, vector fields, tensor fields, etc.

[23] The aptness of this way of writing the action will be apparent from the examples. In the treatment of vibrating systems given in Sec. 36, we did not explicitly introduce the functions $L = T - U$ and $\mathscr{L}$.   Of course, in some cases, e.g., the vibrating plate, $\mathscr{L}$ must involve higher-order derivatives.

where ∇ is the operator

$$\left(\frac{\partial}{\partial x_0}, \frac{\partial}{\partial x_1}, \cdots, \frac{\partial}{\partial x_n}\right),$$

R is some n-dimensional region, and Ω is the "cylindrical space-time region" $R \times [a, b]$, i.e., the Cartesian product of R and the interval $[a, b]$ (see footnote 10, p. 164). The functions $L(u, \nabla u)$ and $\mathscr{L}(u, \nabla u)$ are called the *Lagrangian* and *Lagrangian density* of the field, respectively. Applying the principle of stationary action to (117), we require that $\delta J = 0$. This leads to the Euler equations

$$\frac{\partial \mathscr{L}}{\partial u_j} - \sum_{i=0}^{3} \frac{\partial}{\partial x_i} \frac{\partial \mathscr{L}}{\partial\left(\dfrac{\partial u_j}{\partial x_i}\right)} = 0 \qquad (j = 1, \ldots, m), \tag{118}$$

which are the desired field equations.

Example 1. For the vibrating string with free ends ($\varkappa_1 = \varkappa_2 = 0$), we have $m = n = 1$, and

$$\mathscr{L} = \tfrac{1}{2}(\rho u_t^2 - \tau u_x^2) = \tfrac{1}{2}(\rho u_{x_0}^2 - \tau u_{x_1}^2)$$

[cf. formula (16)].

Example 2. For the vibrating membrane with a free boundary [$\varkappa(s) = 0$] we have $m = 1$, $n = 2$, and

$$\mathscr{L} = \tfrac{1}{2}[\rho u_t^2 - \tau(u_x^2 + u_y^2)] = \tfrac{1}{2}[\rho u_{x_0}^2 - \tau(u_{x_1}^2 + u_{x_2}^2)]$$

[cf. formula (42)].

Example 3. Consider the *Klein-Gordon equation*

$$(\square - M^2)u(x) = 0, \tag{119}$$

describing the scalar field corresponding to uncharged particles of mass M with spin zero (e.g., π^0-mesons). Here, $\square$ denotes the *D'Alembertian (operator)*

$$\square = -\frac{\partial^2}{\partial x_0^2} + \frac{\partial^2}{\partial x_1^2} + \frac{\partial^2}{\partial x_2^2} + \frac{\partial^2}{\partial x_3^2}.$$

It is easy to see that (119) is the Euler equation corresponding to the Lagrangian density

$$\mathscr{L} = \tfrac{1}{2}(u_{x_0}^2 - u_{x_1}^2 - u_{x_2}^2 - u_{x_3}^2 - M^2 u^2). \tag{120}$$

38.2. Conservation laws for fields. Noether's theorem (derived in Sec. 37.5) affords a general method of deriving *conservation laws* for fields, i.e., for constructing combinations of field functions, called *field invariants*, which do not change in time. Thus, suppose the integral

$$\int_\Omega \mathscr{L}(u, \nabla u)\, dx$$

is invariant under an *r*-parameter family of transformations[24]

$$x_i^* = \Phi_i(x, u, \nabla u; \varepsilon) \sim x_i + \sum_{k=1}^r \varepsilon_k \varphi_i^{(k)} \qquad (i = 0, 1, 2, 3),$$

$$u_j^* = \psi_j(x, u, \nabla u; \varepsilon) \sim u_j + \sum_{k=1}^r \varepsilon_k \psi_j^{(k)} \qquad (j = 1, \ldots, m), \qquad (121)$$

where $\varepsilon = (\varepsilon_1, \ldots, \varepsilon_r)$. Then, according to Remark 3, p. 179, we have *r* relations of the form

$$\operatorname{div} I^{(k)} \equiv \sum_{i=0}^3 \frac{\partial I_i^{(k)}}{\partial x_i} = 0,$$

where

$$I_i^{(k)} = \sum_{j=1}^m \frac{\partial \mathscr{L}}{\partial\left(\dfrac{\partial u_j}{\partial x_i}\right)} \bar{\psi}_j^{(k)} + \mathscr{L}\varphi_i^{(k)} \qquad (k = 1, \ldots, r) \qquad (122)$$

and

$$\bar{\psi}_j^{(k)} = \psi_j^{(k)} - \sum_{i=1}^n \frac{\partial u_j}{\partial x_i} \varphi_i^{(k)}.$$

These equations have the following interesting consequence: Suppose the cylinder $\Omega = R \times [a, b]$, where R is the three-dimensional sphere defined by

$$x_1^2 + x_2^2 + x_3^2 \leqslant c^2.$$

Let Γ be the boundary of Ω, and let ν be the unit outward normal to Γ. Then, integrating each of the relations (122) over Γ and using Green's theorem [formula (5) of Sec. 35], we obtain

$$\int_\Omega \operatorname{div} I^{(k)}\, dx = \int_\Gamma (I^{(k)}, \nu)\, d\sigma = 0 \qquad (k = 1, \ldots, r). \qquad (123)$$

The surface integral in (123) is the sum of an integral over the lateral surface of the cylinder Γ and an integral over the two end surfaces cut off by the planes $x_0 = a$, $x_0 = b$. As $c \to \infty$, the integral over the lateral surfaces goes to zero (by the usual argument requiring that the field fall off at infinity "sufficiently rapidly"), and we are left with the integral over the end surfaces.

---

[24] From now on, we set $n = 3$.

On these surfaces, the scalar product $(I^{(k)}, \nu)$ reduces to $I_0^{(k)}$, where the plus sign refers to the "top" surface and the minus sign to the "bottom" surface. Therefore, taking the limit as $c \to \infty$ in (123), we find that

$$\int I_0^{(k)} (a, x_1, x_2, x_3) \, dx_1 \, dx_2 \, dx_3$$
$$= \int I_0^{(k)} (b, x_1, x_2, x_3) \, dx_1 \, dx_2 \, dx_3 \qquad (k = 1, \ldots, r), \tag{124}$$

where $I_0^{(k)}$ denotes the x_0-component of the vector $I^{(k)}$, and the integrations extend over all of three-dimensional space, as will always be assumed if no region of integration is indicated. Since a and b are arbitrary, it follows from (124) that the quantities

$$\int I_0^{(k)} \, dx_1 \, dx_2 \, dx_3$$
$$= \int \left(\sum_{j=1}^{m} \frac{\partial \mathscr{L}}{\partial \left(\dfrac{\partial u_j}{\partial x_0} \right)} \psi_j^{(k)} + \mathscr{L} \varphi_0^{(k)} \right) dx_1 \, dx_2 \, dx_3 \qquad (k = 1, \ldots, r) \tag{125}$$

are independent of time. The r quantities (125) are the required field invariants, whose existence is implied by the invariance of the action functional under the r-parameter family of transformations (121).

Remark. Of course, all the functions in (125) are supposed to be evaluated on an extremal surface of the action functional, corresponding to a solution $u(x)$ of the field equations (118).

38.3. Conservation of energy and momentum. The action functional of any physical field is invariant under parallel displacements, i.e., under the family of transformations

$$\begin{aligned} x_i^* &= x_i + \varepsilon_i \qquad (i = 0, 1, 2, 3), \\ u_j^* &= u_j \qquad\quad (j = 1, \ldots, m), \end{aligned} \tag{126}$$

where the ε_i are arbitrary. In this case, we have

$$\delta x_i = \varepsilon_i, \qquad \delta u_j = 0,$$

which implies

$$\varphi_i^{(k)} = \delta_{ik}, \qquad \psi_j^{(k)} = - \sum_{i=1}^{n} \frac{\partial u_j}{\partial x_i} \delta_{ik} = - \frac{\partial u_j}{\partial x_k},$$

where δ_{ik} is the Kronecker delta. According to (125), the corresponding field invariants are

$$\int \left(\sum_{j=1}^{m} \frac{\partial \mathscr{L}}{\partial \left(\dfrac{\partial u_j}{\partial x_0} \right)} \frac{\partial u_j}{\partial x_k} - \mathscr{L} \delta_{0k} \right) dx_1 \, dx_2 \, dx_3 \qquad (k = 0, 1, 2, 3).$$

It is convenient to introduce the second-rank tensor

$$T_{ik} = \sum_{j=1}^{m} \frac{\partial \mathcal{L}}{\partial \left(\dfrac{\partial u_j}{\partial x_i} \right)} \frac{\partial u_j}{\partial x_k} - \mathcal{L} \delta_{ik}, \qquad (127)$$

called the *energy-momentum tensor*. In terms of T_{ik}, the field invariants are

$$P_k = \int T_{0k} \, dx_1 \, dx_2 \, dx_3 \qquad (k = 0, 1, 2, 3).$$

The vector

$$P = (P_0, P_1, P_2, P_3)$$

is called the *energy-momentum vector*, and in fact, it can be shown that P_0 is the energy and P_1, P_2, P_3 the momentum components of the field. Thus, since P is a field invariant, we have just proved that *the energy and momentum of the field are conserved.*

38.4. Conservation of angular momentum. According to the special theory of relativity, the action functional of any physical field is invariant under *orthochronous Lorentz transformations*, i.e., under transformations of four-dimensional space-time which leave the quadratic form

$$-x_0^2 + x_1^2 + x_2^2 + x_3^2$$

invariant and preserve the time direction.[25] For simplicity, we consider the case where $u(x)$ is a scalar field ($m = 1$). Then the action functional must be invariant under the family of (infinitesimal) transformations

$$\begin{aligned} x_i^* &\sim x_i + \sum_{l \neq i} g_{ll} \varepsilon_{il} x_l, \\ u^* &= u, \end{aligned} \qquad (128)$$

where

$$g_{00} = -1, \qquad g_{11} = g_{22} = g_{33} = 1$$

and

$$\varepsilon_{kl} = -\varepsilon_{lk} \qquad (k \neq l) \qquad (129)$$

are the parameters determining the given transformation.[26] Since the twelve parameters ε_{kl} ($k \neq l$) are connected by the relations (129), only six of them are independent, and we choose the independent parameters to be those for which $k < l$.

---

[25] The determinant of the matrix corresponding to a Lorentz transformation equals ± 1, where the plus sign corresponds to the so-called *proper* Lorentz transformations. See e.g., V. I. Smirnov, *Linear Algebra and Group Theory*, translated by R. A. Silverman, McGraw-Hill Book Co., Inc., New York (1961), Chap. 7.

[26] The parameters $\varepsilon_{12}, \varepsilon_{13}, \varepsilon_{23}$ are angles of rotation, while $\varepsilon_{01}, \varepsilon_{02}, \varepsilon_{03}$ are certain expressions involving the velocity of light and the velocity of one physical reference frame with respect to the other.

Corresponding to the transformations (128), we have

$$
\begin{aligned}
\delta x_i &= \sum_{l \neq i} g_{ll} \varepsilon_{il} x_l = \sum_{l \neq k} \sum_{k=0}^{3} g_{ll} \varepsilon_{kl} \, \delta_{ik} x_l \\
&= \sum_{l < k} \sum_{k=0}^{3} g_{ll} \varepsilon_{kl} \, \delta_{ik} x_l + \sum_{k > l} \sum_{k=0}^{3} g_{ll} \varepsilon_{kl} \, \delta_{ik} x_l \\
&= \sum_{l < k} \sum_{k=0}^{3} \varepsilon_{kl} (g_{ll} \, \delta_{ik} x_l - g_{kk} \, \delta_{il} x_k),
\end{aligned}
$$

where δ_{ik} is the Kronecker delta, and

$$
\overline{\delta u} = - \sum_{i=0}^{3} \frac{\partial u}{\partial x_i} \, \delta x_i.
$$

It follows that

$$
\varphi_i^{(k, \, l)} = g_{ll} \, \delta_{ik} x_l - g_{kk} \, \delta_{il} x_k,
$$
$$
\psi^{(k, \, l)} = \sum_{i=0}^{3} \frac{\partial u}{\partial x_i} (g_{kk} \, \delta_{il} x_k - g_{ll} \, \delta_{ik} x_l) = \frac{\partial u}{\partial x_l} g_{kk} x_k - \frac{\partial u}{\partial x_k} g_{ll} x_l,
$$

where the pair of indices k, l plays the same role as the single index k in (121) and ranges over the six combinations

$$
0, 1; \quad 0, 2; \quad 0, 3; \quad 1, 2; \quad 1, 3; \quad 2, 3.
$$

According to (125), the corresponding field invariants are

$$
\int \left(\frac{\partial \mathscr{L}}{\partial \left(\frac{\partial u}{\partial x_i} \right)} \left[\frac{\partial u}{\partial x_l} g_{kk} x_k - \frac{\partial u}{\partial x_k} g_{ll} x_l \right] \right.
$$

$$
\left. + \mathscr{L} [g_{ll} \, \delta_{ik} x_l - g_{kk} \, \delta_{il} x_k] \right) dx_1 \, dx_2 \, dx_3 \qquad (k < l). \tag{130}
$$

It is convenient to introduce the third-rank tensor

$$
M_{ik} = \frac{\partial \mathscr{L}}{\partial \left(\frac{\partial u}{\partial x_i} \right)} \left[\frac{\partial u}{\partial x_l} g_{kk} x_k - \frac{\partial u}{\partial x_k} g_{ll} x_l \right] + \mathscr{L} [g_{ll} \, \delta_{ik} x_l - g_{kk} \, \delta_{il} x_k] \quad (k < l),
$$
$$
M_{ikl} = - M_{ilk} \qquad (k > l), \tag{131}
$$

called the *angular momentum tensor*. By definition, M_{ikl} is antisymmetric in the indices k and l. Using the expression (127) for the energy-momentum tensor (specialized to the case of scalar fields), we can write (131) as

$$
M_{ikl} = g_{kk} x_k T_{il} - g_{ll} x_l T_{ik}.
$$

In terms of M_{ikl}, the field invariants are

$$
\int M_{0kl} \, dx_1 \, dx_2 \, dx_3 \qquad (k < l),
$$

a fact summarized by saying that *the angular momentum of the field is conserved*.

Example. Using the quantities g_{ii}, we can write the Lagrangian density (120) corresponding to the Klein-Gordon equation in the form

$$\mathscr{L} = -\frac{1}{2} \sum_{i=0}^{3} g_{ii} \left(\frac{\partial u}{\partial x_i}\right)^2 - \frac{1}{2} M^2 u^2.$$

This leads to the energy-momentum tensor

$$T_{ik} = -g_{ii} \frac{\partial u}{\partial x_i} \frac{\partial u}{\partial x_k} - \mathscr{L} \delta_{ik} \tag{132}$$

and the angular momentum tensor

$$M_{ikl} = g_{ii} \frac{\partial u}{\partial x_i} \left(g_{ll} x_l \frac{\partial u}{\partial x_k} - g_{kk} x_k \frac{\partial u}{\partial x_l}\right) + \mathscr{L}(g_{ll} x_l \delta_{ik} - g_{kk} x_k \delta_{il}).$$

The energy density corresponding to (132) is

$$T_{00} = \frac{1}{2} \sum_{i=0}^{3} \left(\frac{\partial u}{\partial x_i}\right)^2 + \frac{1}{2} M^2 u^2,$$

while the momentum density has the components

$$T_{0k} = \frac{\partial u}{\partial x_0} \frac{\partial u}{\partial x_k} \qquad (k = 1, 2, 3).$$

38.5. The electromagnetic field. To illustrate the methods developed above, we now derive the equations of the electromagnetic field from a suitable Lagrangian density. The electromagnetic field is described by two three-dimensional vectors, the *electric field vector* $E = (E_1, E_2, E_3)$ and the *magnetic field vector* $H = (H_1, H_2, H_3)$. In the absence of electric charges, E and H are related by the familiar *Maxwell equations*

$$\begin{aligned} \operatorname{curl} E &= -\frac{\partial H}{\partial x_0}, & \operatorname{curl} H &= \frac{\partial E}{\partial x_0}, \\ \operatorname{div} H &= 0, & \operatorname{div} E &= 0, \end{aligned} \tag{133}$$

where

$$\operatorname{div} E = \frac{\partial E_1}{\partial x_1} + \frac{\partial E_2}{\partial x_2} + \frac{\partial E_3}{\partial x_3},$$

$$\operatorname{curl} E = \left(\frac{\partial E_3}{\partial x_2} - \frac{\partial E_2}{\partial x_3}, \frac{\partial E_1}{\partial x_3} - \frac{\partial E_3}{\partial x_1}, \frac{\partial E_2}{\partial x_1} - \frac{\partial E_1}{\partial x_2}\right),$$

and similarly for div H, curl H. It is convenient to express E and H in terms of a four-dimensional *electromagnetic potential* $\{A_j\} = (A_0, A_1, A_2, A_3)$,[27] by setting

$$E = \operatorname{grad} A_0 - \frac{\partial A}{\partial x_0}, \qquad H = \operatorname{curl} A, \tag{134}$$

---

[27] Since the symbol A is reserved for the three-dimensional vector (A_1, A_2, A_3), we denote the four-dimensional vector (A_0, A_1, A_2, A_3) by $\{A_j\}$. A is sometimes called the *vector potential* and A_0 the *scalar potential*.

where

$$A = (A_1, A_2, A_3)$$

and

$$\text{grad } A_0 = \left(\frac{\partial A_0}{\partial x_1}, \frac{\partial A_0}{\partial x_2}, \frac{\partial A_0}{\partial x_3} \right).$$

The potential $\{A_j\}$ is not uniquely determined by the vectors E and H. In fact, E and H do not change if we make a *gauge transformation*, i.e., if we replace $\{A_j\}$ by a new potential $\{A'_j\}$ with components

$$A'_j(x) = A_j(x) + \frac{\partial f(x)}{\partial x_j} \qquad (j = 0, 1, 2, 3),$$

where $x = (x_0, x_1, x_2, x_3)$ and $f(x)$ is an arbitrary function. To avoid this lack of uniqueness, an extra condition can be imposed on $\{A_j\}$. The condition usually chosen is

$$-\frac{\partial A_0}{\partial x_0} + \text{div } A = \sum_{j=0}^{3} g_{jj} \frac{\partial A_j}{\partial x_j} = 0, \tag{135}$$

and is known as the *Lorentz condition*.

Next, we prove that the Maxwell equations (133) reduce to a single equation determining the electromagnetic potential $\{A_j\}$. First, we introduce the antisymmetric tensor H_{ij}, whose matrix

$$\begin{Vmatrix} 0 & -E_1 & -E_2 & -E_3 \\ E_1 & 0 & H_3 & -H_2 \\ E_2 & -H_3 & 0 & H_1 \\ E_3 & H_2 & -H_1 & 0 \end{Vmatrix}$$

is formed from the components of E and H. It is easily verified that the formula relating H_{ij} to the potential $\{A_j\}$ is

$$H_{ij} = \frac{\partial A_j}{\partial x_i} - \frac{\partial A_i}{\partial x_j}. \tag{136}$$

In terms of the tensor H_{ij}, we can write the Maxwell equations (133) in the form

$$\sum_{i=0}^{3} g_{ii} \frac{\partial H_j}{\partial x_i} = 0 \qquad (j = 0, 1, 2, 3), \tag{137}$$

$$\frac{\partial H_{ij}}{\partial x_k} + \frac{\partial H_{ki}}{\partial x_j} + \frac{\partial H_{jk}}{\partial x_i} = 0, \tag{138}$$

where in (138),

$$i, j, k = \begin{cases} 0, 1, 2, \\ 1, 2, 3, \\ 2, 3, 0, \\ 3, 0, 1. \end{cases}$$

Substituting (136) into (137) and (138), and using the Lorentz condition (135), we find that (138) is an identity, while (137) reduces to

$$\Box A_j = 0 \qquad (j = 0, 1, 2, 3), \tag{139}$$

where $\Box$ is the D'Alembertian

$$\Box = -\frac{\partial^2}{\partial x_0^2} + \frac{\partial^2}{\partial x_1^2} + \frac{\partial^2}{\partial x_2^2} + \frac{\partial^2}{\partial x_3^2}.$$

Finally, we show that (139) is a consequence of the principle of stationary action,[28] if we choose the Lagrangian density of the electromagnetic field to be

$$\mathscr{L} = \frac{1}{8\pi}(E^2 - H^2). \tag{140}$$

Replacing E and H in (140) by their expressions (134) in terms of the electromagnetic potential $\{A_j\}$, we obtain

$$\mathscr{L} = \frac{1}{8\pi}\left[\left(\operatorname{grad} A_0 - \frac{\partial A}{\partial x_0}\right)^2 - (\operatorname{curl} A)^2\right]. \tag{141}$$

We shall only verify that the Euler equations

$$\frac{\partial \mathscr{L}}{\partial A_j} - \sum_{i=0}^{3} \frac{\partial \mathscr{L}}{\partial\left(\frac{\partial A_j}{\partial x_i}\right)} = 0 \qquad (j = 0, 1, 2, 3) \tag{142}$$

corresponding to (141) can be reduced to the form (139) for the component A_0, since the calculations for A_1, A_2, A_3 are completely analogous. It follows from (141) that

$$\frac{\partial \mathscr{L}}{\partial A_0} = \frac{\partial \mathscr{L}}{\partial\left(\frac{\partial A_0}{\partial x_0}\right)} = 0,$$

$$\frac{\partial \mathscr{L}}{\partial\left(\frac{\partial A_0}{\partial x_1}\right)} = \frac{1}{4\pi}\left(\frac{\partial A_0}{\partial x_1} - \frac{\partial A_1}{\partial x_0}\right),$$

$$\frac{\partial \mathscr{L}}{\partial\left(\frac{\partial A_0}{\partial x_2}\right)} = \frac{1}{4\pi}\left(\frac{\partial A_0}{\partial x_2} - \frac{\partial A_2}{\partial x_0}\right),$$

$$\frac{\partial \mathscr{L}}{\partial\left(\frac{\partial A_0}{\partial x_3}\right)} = \frac{1}{4\pi}\left(\frac{\partial A_0}{\partial x_3} - \frac{\partial A_3}{\partial A_0}\right).$$

---

[28] Provided A satisfies the Lorentz condition.

Thus, for $j = 0$, (142) becomes

$$\frac{\partial \mathscr{L}}{\partial A_0} - \sum_{i=0}^{3} \frac{\partial \mathscr{L}}{\partial \left(\frac{\partial A_0}{\partial x_i} \right)}$$

$$= -\frac{1}{4\pi} \left[\frac{\partial^2 A_0}{\partial x_1^2} + \frac{\partial^2 A_0}{\partial x_2^2} + \frac{\partial^2 A_0}{\partial x_3^2} - \frac{\partial}{\partial x_0} \left(\frac{\partial A_1}{\partial x_1} + \frac{\partial A_2}{\partial x_2} + \frac{\partial A_3}{\partial x_3} \right) \right] = 0.$$

(143)

According to the Lorentz condition (135),

$$\frac{\partial A_1}{\partial x_1} + \frac{\partial A_2}{\partial x_2} + \frac{\partial A_3}{\partial x_3} = \frac{\partial A_0}{\partial x_0},$$

and hence (143) reduces to

$$-\frac{\partial^2 A_0}{\partial x_0^2} + \frac{\partial^2 A_0}{\partial x_1^2} + \frac{\partial^2 A_0}{\partial x_2^2} + \frac{\partial^2 A_0}{\partial x_3^2} = \square A_0 = 0,$$

which is just (139), for $j = 0$.

Remark 1. In deriving (139) from (141), we made use of the Lorentz condition (135). Instead, we could have introduced an additional term into the Lagrangian density by writing

$$\mathscr{L} = \frac{1}{8\pi} \left\{ \left(\text{grad } A_0 - \frac{\partial A}{\partial x_0} \right)^2 - (\text{curl } A)^2 - \left(\text{div } A - \frac{\partial A_0}{\partial x_0} \right)^2 \right\}, \quad (144)$$

which reduces to (141) if the Lorentz condition is satisfied. The Euler equations corresponding to (144) reduce to (139) for *arbitrary* $\{A_j\}$.

Remark 2. The Lagrangian density of the electromagnetic field, and hence its action functional, is invariant under parallel displacements, Lorentz transformations and gauge transformations. According to Sec. 38.3, the invariance under parallel displacements implies conservation of energy and momentum of the field, while, according to Sec. 38.4, the invariance under Lorentz transformations implies conservation of angular momentum of the field. Moreover, according to Remark 4, p. 179, the invariance under gauge transformations (which depend on one arbitrary function) implies the existence of a relation between the left-hand sides of the corresponding Euler equations (139). Therefore, these equations do not uniquely determine the electromagnetic potential $\{A_j\}$. In fact, to determine $\{A_j\}$ uniquely, we need an extra equation, which is usually chosen to be the Lorentz condition (135).[29]

---

[29] The Maxwell equations are actually invariant under a 15-parameter family (group) of transformations. In addition to the 10 conservation laws already mentioned (energy, momentum and angular momentum), this invariance leads to 5 more conservation laws, which, however, do not have direct physical meaning. For a detailed treatment of this problem, see E. Bessel-Hagen, *Über die Erhaltungssätze der Elektrodynamik*, Math. Ann., **84**, 258 (1921).

PROBLEMS

1. Find the Euler equation of the functional

$$J[u] = \int \cdots \int_R \sum_{i=1}^{n} u_{x_i}^2 \, dx_1 \ldots dx_n.$$

2. Find the Euler equation of the functional

$$J[u] = \iiint_R \sqrt{1 + u_x^2 + u_y^2 + u_z^2} \, dx \, dy \, dz.$$

3. Write the appropriate generalization of the Euler equation for the functional

$$J[u] = \iint_R F(x, y, u, u_x, u_y, u_{xx}, u_{xy}, u_{yy}) \, dx \, dy.$$

4. Starting from Green's theorem

$$\iint_R \left(\frac{\partial Q}{\partial x} - \frac{\partial P}{\partial y} \right) dx \, dy = \int_\Gamma (P \, dx + Q \, dy),$$

prove that

$$\iint_R \varphi \frac{\partial^2 \psi}{\partial x^2} \, dx \, dy = \iint_R \psi \frac{\partial^2 \varphi}{\partial x^2} \, dx \, dy + \int_\Gamma \left(\varphi \frac{\partial \psi}{\partial x} - \psi \frac{\partial \varphi}{\partial x} \right) dy,$$

$$\iint_R \varphi \frac{\partial^2 \psi}{\partial y^2} \, dx \, dy = \iint_R \psi \frac{\partial^2 \varphi}{\partial y^2} \, dx \, dy - \int_\Gamma \left(\varphi \frac{\partial \psi}{\partial y} - \psi \frac{\partial \varphi}{\partial y} \right) dx,$$

$$\iint_R \varphi \frac{\partial^2 \psi}{\partial x \, \partial y} \, dx \, dy = \iint_R \psi \frac{\partial^2 \varphi}{\partial x \, \partial y} \, dx \, dy - \frac{1}{2} \int_\Gamma \left(\varphi \frac{\partial \psi}{\partial x} - \psi \frac{\partial \varphi}{\partial x} \right) dx$$
$$+ \frac{1}{2} \int_\Gamma \left(\varphi \frac{\partial \psi}{\partial y} - \psi \frac{\partial \varphi}{\partial y} \right) dy.$$

5. Let $J[u]$ be the functional

$$\frac{1}{2} \int_{t_0}^{t_1} \iint_R \left[-(u_{xx} + u_{yy})^2 + 2(1 - \mu)(u_{xx}u_{yy} - u_{xy}^2) \right] dx \, dy \, dt.$$

Using the result of the preceding problem, prove that if we go from u to $u + \varepsilon\psi$, then

$$\delta J = \varepsilon \int_{t_0}^{t_1} \iint_R (-\nabla^4 u)\psi \, dx \, dy \, dt + \varepsilon \int_{t_0}^{t_1} \int_\Gamma \left[P(u)\psi + M(u) \frac{\partial \psi}{\partial u} \right] ds \, dt,$$

where $M(u)$ and $P(u)$ are given by formulas (61) and (62).

Hint. Express $\partial \psi / \partial x$, $\partial \psi / \partial y$ in terms of $\partial \psi / \partial n$, $\partial \psi / \partial s$, and use integration by parts to get rid of $\partial \psi / \partial s$.

6. Show that when $n = 1$, formula (105) of Sec. 37.4 reduces to formula (7) of Sec. 13.

7. Given the functional

$$J[\sigma] = \iint_R \frac{\partial u}{\partial x} \, dx \, dy,$$

compute $J[\sigma^*]$ if σ^* is obtained from σ by the transformation (108).

8. Derive the Euler equations corresponding to the Lagrangian density

$$\mathscr{L} = \sum_{i=0}^{3} \varepsilon_i \left(\frac{\partial u}{\partial x_i} - eA_i \right)^2 + M^2 u^2 + \sum_{i=0}^{3} \sum_{j=0}^{3} \varepsilon_i \varepsilon_j \left(\frac{\partial A_i}{\partial x_j} \right)^2 + M \sum_{i=0}^{3} \varepsilon_i A_i^2,$$

where the field variables are u, A_0, A_1, A_2, A_3, and the factor ε_i equals 1 if $i = 0$ and -1 if $i = 1, 2, 3$.

9. Show that the Lagrangian density $\mathscr{L}$ of the preceding problem is Lorentz-invariant if u transforms like a scalar and if A_0, A_1, A_2, A_3 transform like the components of a vector under Lorentz transformations. Use this fact to derive various conservation laws for the field described by $\mathscr{L}$.

8

DIRECT METHODS
IN THE
CALCULUS OF VARIATIONS

So far, the basic approach used to solve a given variational problem (and indeed, to prove the *existence* of a solution) has been to reduce the problem to one involving a differential equation (or perhaps a system of differential equations). However, this approach is not always effective, and is greatly complicated by the fact that what is needed to solve a given variational problem is not a solution of the corresponding differential equation in a small neighborhood of some point (as is usually the case in the theory of differential equations), but rather a solution in some fixed region R, which satisfies prescribed boundary conditions on the boundary of R. The difficulties inherent in this approach (especially when several independent variables are involved, so that the differential equation is a *partial differential equation*) have led to a search for variational methods of a different kind, known as *direct methods*, which do not entail the reduction of variational problems to problems involving differential equations.

Once they have been developed, direct variational methods can be used to solve differential equations, and this technique, the inverse of the one we have used until now, plays an important role in the modern theory of the subject. The basic idea is the following: Suppose it can be shown that a given differential equation is the Euler equation of some functional, and suppose it has been proved somehow that this functional has an extremum for a sufficiently smooth admissible function. Then, this very fact proves that the differential equation has a solution satisfying the boundary conditions corresponding to the given variational problem. Moreover, as we shall show below (Sec. 41), variational methods can be used not only to prove the existence of a solution of the original differential equation, but also to calculate a solution to any desired accuracy.

39. Minimizing Sequences

There are many different techniques lumped together under the heading of "direct methods." However, the direct methods considered here are all based on the same general idea, which goes as follows:

Consider the problem of finding the minimum of a functional $J[y]$ defined on a space $\mathcal{M}$ of admissible functions y. For the problem to make sense, it must be assumed that there are functions in $\mathcal{M}$ for which $J[y] < +\infty$, and moreover that[1]

$$\inf_y J[y] = \mu > -\infty, \tag{1}$$

where the greatest lower bound is taken over all admissible y. Then, by the definition of μ, there exists an infinite sequence of functions $\{y_n\} = y_1, y_2, \ldots$, called a *minimizing sequence*, such that

$$\lim_{n \to \infty} J[y_n] = \mu.$$

If the sequence $\{y_n\}$ has a limit function $\hat{y}$, and if it is legitimate to write

$$J[\hat{y}] = \lim_{n \to \infty} J[y_n], \tag{2}$$

i.e.,

$$J[\lim_{n \to \infty} y_n] = \lim_{n \to \infty} J[y_n],$$

then

$$J[\hat{y}] = \mu,$$

and $\hat{y}$ is the solution of the variational problem. Moreover, the functions of the minimizing sequence $\{y_n\}$ can be regarded as approximate solutions of our problem.

Thus, to solve a given variational problem by the direct method, we must

1. Construct a minimizing sequence $\{y_n\}$;

2. Prove that $\{y_n\}$ has a limit function $\hat{y}$;

3. Prove the legitimacy of taking the limit (2).

Remark 1. Two direct methods, the *Ritz method* and the *method of finite differences*, each involving the construction of a minimizing sequence, will be discussed in the next section. We reiterate that a minimizing sequence can always be constructed if (1) holds.

---

[1] By inf is meant the *greatest lower bound* or *infimum*.

Remark 2. Even if a minimizing sequence $\{y_n\}$ exists for a given variational problem, it may not have a limit function $\hat{y}$. For example, consider the functional

$$J[y] = \int_{-1}^{1} x^2 y'^2 \, dx,$$

where

$$y(-1) = -1, \qquad y(1) = 1. \tag{3}$$

Obviously, $J[y]$ takes only positive values and

$$\inf_{y} J[y] = 0.$$

We can choose

$$y_n(x) = \frac{\tan^{-1} nx}{\tan^{-1} n} \qquad (n = 1, 2, \ldots) \tag{4}$$

as the minimizing sequence, since

$$\int_{-1}^{1} \frac{n^2 x^2 \, dx}{(\tan^{-1} n)^2 (1 + n^2 x^2)^2} < \frac{1}{(\tan^{-1} n)^2} \int_{-1}^{1} \frac{dx}{1 + n^2 x^2} = \frac{2}{n \tan^{-1} n},$$

and hence $J[y_n] \to 0$ as $n \to \infty$. But as $n \to \infty$, the sequence (4) has no limit in the class of continuous functions satisfying the boundary conditions (3).

Even if the minimizing sequence $\{y_n\}$ has a limit $\hat{y}$ in the sense of the $\mathscr{C}$-norm (i.e., $y_n \to \hat{y}$ as $n \to \infty$, without any assumptions about the convergence of the derivatives of y_n), it is still no trivial matter to justify taking the limit (2), since in general, the functionals considered in the calculus of variations are not continuous in the $\mathscr{C}$-norm. However, (2) still holds if continuity of $J[y]$ is replaced by a weaker condition:

THEOREM. *If $\{y_n\}$ is a minimizing sequence of the functional $J[y]$, with limit function $\hat{y}$, and if $J[y]$ is lower semicontinuous at $\hat{y}$,[2] then*

$$J[\hat{y}] = \lim_{n \to \infty} J[y_n].$$

Proof. On the one hand,

$$J[\hat{y}] \geqslant \lim_{n \to \infty} J[y_n] = \inf J[y], \tag{5}$$

while, on the other hand, given any $\varepsilon > 0$,

$$J[y_n] - J[\hat{y}] > -\varepsilon, \tag{6}$$

if n is sufficiently large. Letting $n \to \infty$ in (6), we obtain

$$J[\hat{y}] \leqslant \lim_{n \to \infty} J[y_n] + \varepsilon,$$

or

$$J[\hat{y}] \leqslant \lim_{n \to \infty} J[y_n], \tag{7}$$

---

[2] See Remark 1, p. 7.

since ε is arbitrary.   Comparing (5) and (7), we find that

$$J[\hat{y}] = \lim_{n \to \infty} J[y_n],$$

as asserted.

40. The Ritz Method and the Method of Finite Differences[3]

40.1. First, we describe the *Ritz method*, one of the most widely used direct variational methods.   Suppose we are looking for the minimum of a functional $J[y]$ defined on some space $\mathcal{M}$ of admissible functions, which for simplicity we take to be a normed linear space.   Let

$$\varphi_1, \quad \varphi_2, \ldots \tag{8}$$

be an infinite sequence of functions in $\mathcal{M}$, and let $\mathcal{M}_n$ be the n-dimensional linear subspace of $\mathcal{M}$ spanned by the first n of the functions (8), i.e., the set of all linear combinations of the form

$$\alpha_1\varphi_1 + \cdots + \alpha_n\varphi_n, \tag{9}$$

where $\alpha_1, \ldots, \alpha_n$ are arbitrary real numbers.   Then, on each subspace $\mathcal{M}_n$, the functional $J[y]$ leads to a function

$$J[\alpha_1\varphi_1 + \cdots + \alpha_n\varphi_n] \tag{10}$$

of the n variables $\alpha_1, \ldots, \alpha_n$.

Next, we choose $\alpha_1, \ldots, \alpha_n$ in such a way as to minimize (10), denoting the minimum by μ_n and the element of $\mathcal{M}_n$ which yields the minimum by y_n. (In principle, this is a much simpler problem than finding the minimum of the functional $J[y]$ itself.)   Clearly, μ_n cannot increase with n, i.e.,

$$\mu_1 \geqslant \mu_2 \geqslant \cdots,$$

since any linear combination of $\varphi_1, \ldots, \varphi_n$ is automatically a linear combination $\varphi_1, \ldots, \varphi_n, \varphi_{n+1}$.   Correspondingly, each subspace of the sequence

$$\mathcal{M}_1, \quad \mathcal{M}_2, \ldots$$

is contained in the next.   We now give conditions which guarantee that the sequence $\{y_n\}$ is a minimizing sequence.

DEFINITION.   *The sequence* (8) *is said to be complete (in $\mathcal{M}$) if given any $y \in \mathcal{M}$ and any $\varepsilon > 0$, there is a linear combination η_n of the form* (9) *such that $\|\eta_n - y\| < \varepsilon$ (where n depends on ε).*

---

[3] Here we merely outline these two methods, without worrying about questions of convergence, and taking for granted the existence of an exact solution of the given variational problem.

THEOREM.   *If the functional $J[y]$ is continuous,[4] and if the sequence (8) is complete, then*

$$\lim_{n \to \infty} \mu_n = \mu,$$

where

$$\mu = \inf_y J[y].$$

Proof.   Given any $\varepsilon > 0$, let y^* be such that

$$J[y^*] < \mu + \varepsilon.$$

(Such a y^* exists for any $\varepsilon > 0$, by the definition of μ.)   Since $J[y]$ is continuous,

$$|J[y] - J[y^*]| < \varepsilon, \tag{11}$$

provided that $\|y - y^*\| < \delta = \delta(\varepsilon)$.   Let η_n be a linear combination of the form (9) such that $\|\eta_n - y^*\| < \delta$.   (Such an η_n exists for sufficiently large n, since $\{\varphi_n\}$ is complete.)   Moreover, let y_n be the linear combination of the form (9) for which (10) achieves its minimum. Then, using (11), we find that

$$\mu \leqslant J[y_n] \leqslant J[\eta_n] < \mu + 2\varepsilon.$$

Since ε is arbitrary, it follows that

$$\lim_{n \to \infty} J[y_n] = \lim_{n \to \infty} \mu_n = \mu,$$

as asserted.

Remark 1.   The geometric idea of the proof is the following: If $\{\varphi_n\}$ is complete, then any element in the infinite-dimensional space $\mathcal{M}$ can be approximated arbitrarily closely by an element in the finite-dimensional space $\mathcal{M}_n$ (for large enough n).   We can summarize this fact by writing

$$\lim_{n \to \infty} \mathcal{M}_n = \mathcal{M}.$$

Let $\hat{y}$ be the element in $\mathcal{M}$ for which $J[\hat{y}] = \mu$, and let $\hat{y}_n \in \mathcal{M}_n$ be a sequence of functions converging to $\hat{y}$.   Then $\{\hat{y}_n\}$ is a minimizing sequence, since $J[y]$ is continuous.   Although this minimizing sequence cannot be constructed without prior knowledge of $\hat{y}$, we can show that our explicitly constructed sequence $\{y_n\}$ takes values $J[y_n]$ arbitrarily close to $J[\hat{y}_n]$, and hence is itself a minimizing sequence.

Remark 2.   The speed of convergence of the Ritz method for a given variational problem obviously depends both on the problem itself and on

---

[4] I.e., continuous in the norm of $\mathcal{M}$.   For example, functionals of the form

$$J[y] = \int_a^b F(x, y, y') \, dx$$

are continuous in the norm of the space $\mathcal{D}_1(a, b)$.

the choice of the functions φ_n. However, it should be pointed out that in many cases, linear combinations involving only a very small number of functions φ_n are enough to give a quite satisfactory approximation to the exact solution.

Remark 3. More generally, the spaces $\mathcal{M}$ and $\mathcal{M}_n$ need not be normed linear spaces themselves, but only suitable sets of admissible functions belonging to an underlying normed linear space $\mathcal{R}$ (see Remark 3, p. 8). For example, the admissible functions may satisfy boundary conditions like

$$y(a) = A, \qquad y(b) = B$$

(see Sec. 40.2), or a subsidiary condition like

$$\int_a^b y^2(x)\, dx = 1$$

(see Sec. 41). This case can be handled by appropriate modifications of the present method.

40.2. We now describe another method involving a sequence of finite-dimensional approximations to the space $\mathcal{M}$. This is the *method of finite differences*, which has already been encountered in Sec. 7. There, in connection with the derivation of Euler's equation, we noted that the problem of finding an extremum of the functional[5]

$$J[y] = \int_a^b F(x, y, y')\, dx, \qquad y(a) = A, \quad y(b) = B, \tag{12}$$

can be approximated by the problem of finding an extremum of a function of n variables, obtained as follows: We divide the interval $[a, b]$ into $n + 1$ equal subintervals by introducing the points

$$x_0 = a, \quad x_1, \ldots, x_n, \quad x_{n+1} = b, \qquad x_{i+1} - x_i = \Delta x,$$

and we replace the function $y(x)$ by the polygonal line with vertices

$$(x_0, y_0), (x_1, y_1), \ldots, (x_n, y_n), (x_{n+1}, y_{n+1}),$$

where now $y_i = y(x_i)$. Then (12) can be approximated by the sum

$$J(y_1, \ldots, y_n) = \sum_{i=0}^n F\left[x_i, y_i, \frac{y_{i+1} - y_i}{\Delta x}\right] \Delta x, \tag{13}$$

which is a function of n variables. (Recall that $y_0 = A$ and $y_{n+1} = B$ are fixed.) If for each n, we find the polygonal line minimizing (13), we obtain a sequence of approximate solutions to the original variational problem.

---

[5] Here, $\mathcal{M}$ will be a *linear* space only if $A = B = 0$ (cf. Remark 3).

41. The Sturm-Liouville Problem

In this section, we illustrate the application of direct variational methods to differential equations (cf. the remarks on p. 192), by studying the following boundary value problem, known as the *Sturm-Liouville problem*: Let $P = P(x) > 0$ and $Q = Q(x)$ be two given functions, where Q is continuous and P is continuously differentiable, and consider the differential equation

$$-(Py')' + Qy = \lambda y \tag{14}$$

(known as the *Sturm-Liouville equation*), subject to the boundary conditions

$$y(a) = 0, \qquad y(b) = 0. \tag{15}$$

It is required to find the *eigenfunctions* and *eigenvalues* of the given boundary value problem, i.e., the nontrivial solutions[6] of (14), (15) and the corresponding values of the parameter λ.

> THEOREM. *The Sturm-Liouville problem* (14), (15) *has an infinite sequence of eigenvalues* $\lambda^{(1)}, \lambda^{(2)}, \ldots,$ *and to each eigenvalue* $\lambda^{(n)}$ *there corresponds an eigenfunction* $y^{(n)}$ *which is unique to within a constant factor.*

The proof of this theorem will be carried out in stages, and at the same time we shall derive a method for approximating the eigenvalues $\lambda^{(n)}$ and eigenfunctions $y^{(n)}$.

41.1. We begin by observing that (14) is the Euler equation corresponding to the problem of finding an extremum of the quadratic functional

$$J[y] = \int_a^b (Py'^2 + Qy^2)\, dx, \tag{16}$$

subject to the boundary conditions (15) and the subsidiary condition[7]

$$\int_a^b y^2\, dx = 1. \tag{17}$$

Thus, if $y(x)$ is a solution of this variational problem, it is also a solution of the differential equation (14), satisfying the boundary conditions (15). Moreover, $y(x)$ is not identically zero, because of the condition (17).

Next, we apply the Ritz method (see Sec. 40.1) to the functional (16), first

---

[6] In other words, the solutions which are not identically zero. For any value of λ, (14) and (15) are trivially satisfied by the function $y(x) \equiv 0$.

[7] Use the theorem on p. 43, changing λ to $-\lambda$.

verifying that it is bounded from below, as required [cf. formula (1)]. Since $P(x) > 0$, this fact follows from the inequality

$$\int_a^b (Py'^2 + Qy^2)\, dx > \int_a^b Qy^2\, dx \geqslant M \int_a^b y^2\, dx = M,$$

where

$$M = \min_{a \leqslant x \leqslant b} Q(x).$$

For simplicity, we assume that $a = 0$, $b = \pi$, and we choose $\{\sin nx\}$ as the complete sequence of functions $\{\varphi_n(x)\}$ used in the Ritz method. This sequence also has the desirable feature of being *orthogonal*, i.e.,

$$\int_0^\pi \sin kx \sin lx\, dx = 0 \qquad (k \neq l).$$

If a linear combination

$$\sum_{k=1}^n \alpha_k \sin kx \tag{18}$$

is to be admissible, it must satisfy the conditions (15) and (17). The condition (15) is automatically satisfied by our choice of the functions $\sin nx$, but (17) leads to the requirement

$$\int_0^\pi \left(\sum_{k=1}^n \alpha_k \sin kx \right)^2 dx = \frac{\pi}{2} \sum_{k=1}^n \alpha_k^2 = 1. \tag{19}$$

Moreover, for a linear combination (18), the functional $J[y]$ reduces to

$$J_n(\alpha_1, \ldots, \alpha_n) = \int_0^\pi \left[P(x) \left(\sum_{k=1}^n \alpha_k \sin kx \right)'^2 + Q(x) \left(\sum_{k=1}^n \alpha_k \sin kx \right)^2 \right] dx, \tag{20}$$

which is a function of the n variables $\alpha_1, \ldots, \alpha_n$ (in fact, a quadratic form in these variables).

Thus, in terms of the variables $\alpha_1, \ldots, \alpha_n$, our problem is to minimize $J_n(\alpha_1, \ldots, \alpha_n)$ on the surface σ_n of the n-dimensional sphere with equation (19). Since σ_n is a compact set and $J_n(\alpha_1, \ldots, \alpha_n)$ is continuous on σ_n, $J_n(\alpha_1, \ldots, \alpha_n)$ has a minimum $\lambda_n^{(1)}$ at some point $\alpha_1^{(1)}, \ldots, \alpha_n^{(1)}$ of σ_n.[8] Let

$$y_n^{(1)}(x) = \sum_{k=1}^n \alpha_k^{(1)} \sin kx$$

be the linear combination (18) achieving the minimum $\lambda_n^{(1)}$. If this procedure is carried out for $n = 1, 2, \ldots$, we obtain a sequence of numbers

$$\lambda_1^{(1)}, \lambda_2^{(1)}, \ldots, \tag{21}$$

and a corresponding sequence of functions

$$y_1^{(1)}(x), y_2^{(1)}(x), \ldots \tag{22}$$

---

[8] See e.g., T. M. Apostol, *op. cit.*, Theorem 4–20, p. 73.

Noting that σ_n is the subset of σ_{n+1} obtained by setting $\alpha_{n+1} = 0$, while

$$J_n(\alpha_1, \ldots, \alpha_n) = J_{n+1}(\alpha_1, \ldots, \alpha_n, 0),$$

we see that

$$\lambda_{n+1}^{(1)} \leqslant \lambda_n^{(1)}, \tag{23}$$

since increasing the domain of definition of a function can only decrease its minimum. It follows from (23) and the fact that $J[y]$ is bounded from below that the limit

$$\lambda^{(1)} = \lim_{n \to \infty} \lambda_n^{(1)} \tag{24}$$

exists.

41.2. Now that we have proved the convergence of the sequence of numbers (21), representing the minima of the functional

$$\int_0^\pi (Py'^2 + Qy^2)\, dx$$

on the sets of functions of the form

$$\sum_{k=1}^n \alpha_k \sin kx$$

satisfying the condition (19), it is natural to try to prove the convergence of the sequence of functions (22) for which these minima are achieved. We first prove a weaker result:

LEMMA 1. *The sequence $\{y_n^{(1)}(x)\}$ contains a uniformly convergent subsequence.*

Proof. For simplicity, we temporarily write $y_n(x)$ instead of $y_n^{(1)}(x)$. The sequence

$$\lambda_n^{(1)} = \int_0^\pi (Py_n'^2 + Qy_n^2)\, dx$$

is convergent and hence bounded, i.e.,

$$\int_0^\pi (Py_n'^2 + Qy_n^2)\, dx \leqslant M$$

for all n, where M is some constant. Therefore

$$\int_0^\pi Py_n'^2\, dx \leqslant M + \left| \int_0^\pi Qy_n^2\, dx \right| \leqslant M + \max_{a \leqslant x \leqslant b} |Q(x)| = M_1,$$

and since $P(x) > 0$,

$$\int_0^\pi y_n'^2(x)\, dx \leqslant \frac{M_1}{\min_{a \leqslant x \leqslant b} P(x)} = M_2. \tag{25}$$

Using (25), the condition

$$y_n(0) = 0,$$

and Schwarz's inequality, we find that

$$|y_n(x)|^2 = \left| \int_0^x y_n'(\xi)\, d\xi \right|^2 \leqslant \int_0^x y_n'^2(\xi)\, d\xi \int_0^x d\xi \leqslant M_2\pi,$$

so that $\{y_n(x)\}$ is uniformly bounded.[9]   Moreover, again using Schwarz's inequality, we have

$$|y_n(x_2) - y_n(x_1)|^2 = \left| \int_{x_1}^{x_2} y_n'(x)\, dx \right|^2 \leqslant \int_{x_1}^{x_2} y_n'^2\, dx \cdot \left| \int_{x_1}^{x_2} dx \right| \leqslant M_2 |x_2 - x_1|,$$

so that $\{y_n(x)\}$ is equicontinuous.[10] Thus, according to Arzelà's theorem,[11] we can select a uniformly convergent subsequence $\{y_{n_m}(x)\}$ from the sequence $\{y_n(x)\}$ and Lemma 1 is proved.

We now set

$$y^{(1)}(x) = \lim_{m \to \infty} y_{n_m}(x). \tag{26}$$

Our object is to show that $y^{(1)}(x)$ satisfies the Sturm-Liouville equation (14) with $\lambda = \lambda^{(1)}$. However, we are still not in a position to take the limit as $m \to \infty$ of the integral

$$\int_0^\pi (P y_{n_m}'^2 + Q y_{n_m}^2)\, dx,$$

since as yet we know nothing about the convergence of the derivatives y_{n_m}'. Therefore, the fact that for each m, the function y_{n_m} minimizes the functional $J[y]$ for y in the n_m-dimensional space spanned by the linear combinations

$$\sum_{k=1}^{n_m} \alpha_k \sin kx$$

[subject to the condition (19) with $n = n_m$] still does not imply that the limit function $y^{(1)}(x)$ minimizes $J[y]$ for y in the full space of admissible functions. To avoid this difficulty, we argue as follows:

LEMMA 2. Let $y(x)$ be continuous in $[0, \pi]$, and let

$$\int_0^\pi [-(Ph')' + Q_1 h] y\, dx = 0 \tag{27}$$

---

[9] A family of functions Ψ defined on $[a, b]$ is said to be *uniformly bounded* if there is a constant M such that

$$|\psi(x)| \leqslant M$$

for all $\psi \in \Psi$ and all $a \leqslant x \leqslant b$.

[10] A family of functions Ψ defined on $[a, b]$ is said to be *equicontinuous* if given any $\varepsilon > 0$, there is a $\delta > 0$ such that

$$|\psi(x_2) - \psi(x_1)| < \varepsilon$$

for all $\psi \in \Psi$, provided that $|x_2 - x_1| < \delta$.

[11] Arzelà's theorem states that every uniformly bounded and equicontinuous sequence of functions contains a uniformly convergent subsequence (converging to a continuous limit function).   See e.g., R. Courant and D. Hilbert, *op. cit.*, vol. 1, p. 59.

for every function $h(x) \in \mathscr{D}_2(0, \pi)$,[12] *satisfying the boundary conditions*

$$h(0) = h(\pi) = 0, \qquad h'(0) = h'(\pi) = 0. \tag{28}$$

Then $y(x)$ also belongs to $\mathscr{D}_2(0, \pi)$, and

$$-(Py')' + Q_1 y = 0.$$

Proof. If we integrate (27) by parts and use (28), we find that

$$\int_0^\pi [-(Ph')' + Q_1 h] \, y \, dx = -\int_0^\pi Ph'' y \, dx - \int_0^\pi P'h'y \, dx + \int_0^\pi Q_1 h y \, dx$$

$$= -\int_0^\pi \left[-Py + \int_0^x P'y \, d\xi + \int_0^x \left(\int_0^\xi Q_1 y \, dt \right) d\xi \right] dx = 0.$$

It follows from Lemma 3, p. 10 that

$$-Py + \int_0^x P'y \, d\xi + \int_0^x \left(\int_0^\xi Q_1 y \, dt \right) d\xi = c_0 + c_1 x, \tag{29}$$

where c_0 and c_1 are constants. Since the right-hand side and the second and third terms in the left-hand side of (29) are obviously differentiable, $(Py)'$ exists, and in fact, differentiating (29) term by term, we find that

$$-(Py)' + P'y + \int_0^x Q_1 y \, d\xi = c_1. \tag{30}$$

Since the function P is continuously differentiable and does not vanish, y' exists and is continuous. Thus, (30) reduces to

$$-Py' + \int_0^x Q_1 y \, d\xi = c_1. \tag{31}$$

Since the right-hand side and the second term in the left-hand side of (31) are differentiable, it follows that $(Py')'$ exists, and in fact

$$-(Py')' + Q_1 y = 0,$$

as asserted. Moreover, by the same argument as before, y'' exists and is continuous.

41.3. We can now show that the function $y^{(1)}(x)$ defined by (26), whose existence follows from Lemma 1, satisfies the Sturm-Liouville equation

$$-(Py^{(1)\prime})' + Qy^{(1)} = \lambda^{(1)} y^{(1)}, \tag{32}$$

where $\lambda^{(1)}$ is the limit (24). According to the theory of Lagrange multipliers (cf. footnote 7, p. 43), at the point $(\alpha_1^{(1)}, \ldots, \alpha_n^{(1)})$ where the quadratic form (20) achieves its minimum subject to the subsidiary condition (19), we have

$$\frac{\partial}{\partial \alpha_r} \left\{ J_n(\alpha_1, \ldots, \alpha_n) - \lambda_n^{(1)} \int_0^\pi \left(\sum_{k=1}^n \alpha_k \sin kx \right)^2 dx \right\} = 0 \qquad (r = 1, \ldots, n).$$

---

[12] I.e., for every $h(x)$ with continuous first and second derivatives in $[0, \pi]$.

This leads to the n equations

$$\int_0^\pi \left\{ P(x) \left[\sum_{k=1}^n \alpha_k^{(1)} (\sin kx)' \right] (\sin rx)' \right.$$
$$\left. + [Q(x) - \lambda_n^{(1)}] \left[\sum_{k=1}^n \alpha_k^{(1)} \sin kx \right] \sin rx \right\} dx = 0 \qquad (r = 1, \ldots, n). \tag{33}$$

Multiplying each of the equations (33) by an arbitrary constant $C_r^{(n)}$ and summing over r from 1 to n, we obtain

$$\int_0^\pi [Py_n' h_n' + (Q - \lambda_n^{(1)}) y_n h_n] \, dx = 0, \tag{34}$$

where

$$h_n(x) = \sum_{r=1}^n C_r^{(n)} \sin rx. \tag{35}$$

An integration by parts transforms (34) into

$$\int_0^\pi [-(Ph_n')' + (Q - \lambda_n^{(1)}) h_n] y_n \, dx = 0. \tag{36}$$

If $h(x)$ is an arbitrary function in $\mathscr{D}_2(0, \pi)$ satisfying the boundary conditions (28), we can choose the coefficients $C_r^{(n)}$ in such a way that

$$h_n \Rightarrow h, \qquad h_n' \Rightarrow h', \qquad h_n'' \Rightarrow h''$$

(see Prob. 8). Here, the symbol $\Rightarrow$ denotes convergence in the mean, i.e., $h_n \Rightarrow h$ stands for

$$\lim_{n \to \infty} \int_0^\pi |h_n(x) - h(x)|^2 \, dx = 0$$

Since $y_n^{(1)} \to y^{(1)}$ uniformly in $[0, \pi]$,[13] it follows from (36) that

$$\lim_{m \to \infty} \int_0^\pi [-(Ph_{n_m}')' + (Q - \lambda_{n_m}^{(1)}) h_{n_m}] y_{n_m}^{(1)} \, dx$$
$$= \int_0^\pi [-(Ph')' + (Q - \lambda^{(1)}) h] y^{(1)} \, dx = 0$$

(see Prob. 9). The fact that $y^{(1)}$ is an element of $\mathscr{D}_2(0, \pi)$ and satisfies the Sturm-Liouville equation (32) is now an immediate consequence of Lemma 2, with $Q_1 = Q - \lambda^{(1)}$.

So far, the function $y^{(1)}(x)$ has been defined as the limit of a subsequence $\{y_{n_m}^{(1)}(x)\}$ of the original sequence $\{y_n^{(1)}(x)\}$. We now show that the sequence

---

[13] We now restore the superscript on $y_n^{(1)}$.

$\{y_n^{(1)}(x)\}$ itself converges to $y^{(1)}(x)$. To prove this, we use the fact that for a given λ, the solution of the Sturm-Liouville equation

$$-(Py')' + Qy = \lambda y \tag{37}$$

satisfying the boundary conditions

$$y(0) = 0, \qquad y(\pi) = 0 \tag{38}$$

and the normalization condition

$$\int_0^\pi y^2(x)\, dx = 1 \tag{39}$$

is unique except for sign. Let $y^{(1)}(x)$ be a solution of (37) corresponding to $\lambda = \lambda^{(1)}$, and suppose $y^{(1)}(x_0) \neq 0$ at some point x_0 in $[0, \pi]$. Then choose the sign so that $y^{(1)}(x_0) > 0$. Similarly, let $y_n^{(1)}(x)$ be a solution of (37) corresponding to $\lambda = \lambda_n^{(1)}$, and choose the signs so that $y_n^{(1)}(x_0) \geqslant 0$ for all n. If $y_n^{(1)}(x)$ does not converge to $y^{(1)}(x)$, we can select another subsequence from $\{y_n^{(1)}(x)\}$ converging to another solution $\bar{y}^{(1)}(x)$ of (37), where again $\lambda = \lambda^{(1)}$. Because of the uniqueness (except for sign) of solutions of (37), subject to (38) and (39), this means that

$$\bar{y}^{(1)}(x) = -y^{(1)}(x),$$

and hence $\bar{y}^{(1)}(x_0) < 0$, which is impossible, since $y_n^{(1)}(x_0) \geqslant 0$ for all n. Therefore, $y_n^{(1)}(x) \to y^{(1)}(x)$ [in fact, uniformly], provided we choose each $y_n^{(1)}(x)$ with the proper sign.

41.4. We have just proved that the Sturm-Liouville problem has the eigenfunction $y^{(1)}(x)$, corresponding to the eigenvalue $\lambda^{(1)}$. The "next" eigenfunction $y^{(2)}(x)$ and the corresponding eigenvalue $\lambda^{(2)}$ can be found by minimizing the quadratic functional

$$J[y] = \int_0^\pi (Py'^2 + Qy^2)\, dx \tag{40}$$

subject to the same conditions (38) and (39) as before, plus an extra orthogonality condition

$$\int_0^\pi y^{(1)}(x) y(x)\, dx = 0. \tag{41}$$

In fact, substituting

$$y(x) = \sum_{k=1}^n \alpha_k \sin kx \tag{42}$$

into (40), we again obtain the quadratic form $J_n(\alpha_1, \ldots, \alpha_n)$ given by (20), but this time we study $J_n(\alpha_1, \ldots, \alpha_n)$ on the set of functions of the form (42) which not only lie on the n-dimensional sphere σ_n with equation (19), thereby satisfying the normalization condition (39), but are also orthogonal to the function

$$y_n^{(1)}(x) = \sum_{k=1}^n \alpha_k^{(1)} \sin kx,$$

i.e., satisfy the condition

$$\sum_{k=1}^{n} \alpha_k \int_0^\pi \sin kx \left(\sum_{l=1}^{n} \alpha_l^{(1)} \sin lx \right) dx = \frac{\pi}{2} \sum_{k=1}^{n} \alpha_k \alpha_k^{(1)} = 0. \tag{43}$$

This is the equation of an $(n-1)$-dimensional hyperplane, passing through the origin of coordinates in n dimensions. Its intersection with the sphere (19) is an $(n-1)$-dimensional sphere $\hat{\sigma}_{n-1}$. By the same argument as before (cf. footnote 8), $J_n(\alpha_1, \ldots, \alpha_n)$ has a minimum $\lambda_n^{(2)}$ on $\hat{\sigma}_{n-1}$. It is not hard to see that

$$\lambda_{n+1}^{(2)} \leqslant \lambda_n^{(2)}$$

[cf. (23)], and hence the limit

$$\lambda^{(2)} = \lim_{n \to \infty} \lambda_n^{(2)}$$

exists, since $J[y]$ is bounded from below. Moreover, it is obvious that

$$\lambda^{(1)} \leqslant \lambda^{(2)}. \tag{44}$$

Now let

$$y_n^{(2)} = \sum_{k=1}^{n} \alpha_k^{(2)} \sin kx$$

be the linear combination (42) achieving the minimum $\lambda_n^{(2)}$, where, of course, the point $(\alpha_1^{(2)}, \ldots, \alpha_n^{(2)})$ lies on the sphere $\hat{\sigma}_{n-1}$. As before, we can show that the sequence $\{y_n^{(2)}(x)\}$ converges uniformly to a limit function $y^{(2)}(x)$ which satisfies the Sturm-Liouville equation (37) [with $\lambda = \lambda^{(2)}$], the boundary conditions (38), the normalization condition (39), and the orthogonality condition (41). In other words, $y^{(2)}(x)$ is the eigenfunction of the Sturm-Liouville problem corresponding to the eigenvalue $\lambda^{(2)}$. Since orthogonal functions cannot be linearly dependent, and since only one eigenfunction corresponds to each eigenvalue (except for a constant factor), we have the strict inequality

$$\lambda^{(1)} < \lambda^{(2)},$$

instead of (44). Finally, we note that by repeating the above argument, with obvious modifications, we can obtain further eigenvalues $\lambda^{(3)}, \lambda^{(4)}, \ldots$, and corresponding eigenfunctions $y^{(3)}(x), y^{(4)}(x), \ldots$.

For further material on the use of direct methods in the calculus of variations, we refer the reader to the abundant literature on the subject.[14]

---

[14] See e.g., N. Krylov, *Les méthodes de solution approchée des problèmes de la physique mathématique*, Mémorial des Sciences Mathématiques, fascicule 49, Gauthier-Villars et Cie., Paris (1931); S. G. Mikhlin, Прямые Методы в Математической Физике (*Direct Methods in Mathematical Physics*), Gos. Izd. Tekh.-Teor. Lit., Moscow (1950); S. G. Mikhlin, Вариационные Методы в Математической Физике (*Variational Methods in Mathematical Physics*), Gos. Izd. Tekh.-Teor. Lit., Moscow (1957); L. V. Kantorovich and V. I. Krylov, *Approximate Methods of Higher Analysis*, translated by C. D. Benster, Interscience Publishers, Inc., New York (1958).

PROBLEMS

1. Let the functional $J[y]$ be such that $J[y] > -\infty$ for some admissible function, and let

$$\sup J[y] = \mu < +\infty,$$

where sup denotes the *least upper bound* or *supremum*. By analogy with the treatment given in Sec. 39, define a *maximizing sequence*, and then state and prove the corresponding version of the theorem on p. 194.

2. Use the Ritz method to find an approximate solution of the problem of minimizing the functional

$$J[y] = \int_0^1 (y'^2 - y^2 - 2xy)\, dx, \qquad y(0) = y(1) = 0,$$

and compare the answer with the exact solution.

Hint. Choose the sequence $\{\varphi_n(x)\}$ (see p. 195) to be

$$x(1 - x), \quad x^2(1 - x), \quad x^3(1 - x), \ldots$$

3. Use the Ritz method to find an approximate solution of the extremum problem associated with the functional

$$J[y] = \int_0^1 (x^3 y''^2 + 100xy^2 - 20xy)\, dx, \qquad y(1) = y'(1) = 0.$$

Hint. Choose the sequence $\{\varphi_n(x)\}$ to be

$$(x - 1)^2, \quad x(x - 1)^2, \quad x^2(x - 1)^2, \ldots$$

4. Use the Ritz method to find an approximate solution of the problem of minimizing the functional

$$J[y] = \int_0^2 (y'^2 + y^2 + 2xy)\, dx, \qquad y(0) = y(2) = 0,$$

and compare the answer with the exact solution.

5. Use the Ritz method to find an approximate solution of the equation

$$\frac{\partial^2 u}{\partial x^2} + \frac{\partial^2 u}{\partial y^2} = -1$$

inside the square

$$R: \qquad -a \leqslant x \leqslant a, \qquad -a \leqslant y \leqslant a,$$

where u vanishes on the boundary of R.

Hint. Study the functional

$$J[u] = \iint_R \left[\left(\frac{\partial u}{\partial x}\right)^2 + \left(\frac{\partial u}{\partial y}\right)^2 - 2u \right] dx\, dy,$$

and choose the two-dimensional generalization of the sequence $\{\varphi_n(x)\}$ to be

$$(x^2 - a^2)(y^2 - b^2), \quad (x^2 + y^2)(x^2 - a^2)(y^2 - b^2), \ldots$$

6. Write the Sturm-Liouville equation associated with the quadratic functional

$$J[y] = \int_a^b (c_1 y'^2 + c y^2) \, dx,$$

where c and $c_1 > 0$ are constants, subject to the boundary conditions

$$y(a) = 0, \qquad y(b) = 0.$$

Find the corresponding eigenvalues and eigenfunctions.

7. Formulate a variational problem leading to the Sturm-Liouville equation (14) subject to the boundary conditions

$$y'(a) = 0, \qquad y'(b) = 0,$$

instead of the boundary conditions (15).

Hint. Recall the natural boundary conditions (29) of Sec. 6.

8. Prove that any function $h(x) \in \mathscr{D}_2(0, \pi)$ satisfying the boundary conditions (28) can be approximated in the mean by a linear combination

$$h_n(x) = \sum_{r=1}^n C_r^{(n)} \sin rx,$$

where at the same time $h_n'(x)$ approximates $h'(x)$ and $h_n''(x)$ approximates $h''(x)$ [in the mean]. Show that the coefficients $C_r^{(n)}$ need not depend on n and can be written simply as C_r.

Hint. Form the Fourier sine series of $h''(x)$ and integrate it twice term by term.

9. Show that if $f_n(x) \to f(x)$ in the mean and $g_n(x) \to g(x)$ uniformly in some interval $[a, b]$, then

$$\int_a^b f_n(x) g_n(x) \, dx \to \int_a^b f(x) g(x) \, dx.$$

Hint. Use Schwarz's inequality.

PROPAGATION OF DISTURBANCES
AND THE
CANONICAL EQUATIONS[1]

In this appendix, we consider the propagation of "disturbances" in a medium which is regarded as being both inhomogeneous and anisotropic. Thus, in general, the velocity of propagation of a disturbance at a given point of the medium will depend both on the position of the point and on the direction of propagation of the disturbance. We also make the following two assumptions about the process under consideration:

1. Each point can be in only one of two states, *excitation* or *rest*, i.e., no concept of the intensity of the disturbance is introduced.

2. If a disturbance arrives at the point P at the time t, then starting from the time t, the point P itself serves as a source of further disturbances propagating in the medium.

In the analysis given here, our aim is to show that a study of processes of excitation of the kind described, together with purely geometric considerations, can be used to derive such basic concepts of the calculus of variations as the canonical equations, the Hamiltonian function, the Hamilton-Jacobi equation, etc. The treatment given here does not rely upon the derivations of these concepts given in the main body of the book (see Secs. 16, 23), and in fact can be used to replace the previous derivations. The reader acquainted

---

[1] The authors would like to acknowledge discussions with M. L. Tsetlyn on the material presented here.

with optics will recognize that we are essentially constructing a mathematical model of the familiar *Huygens' principle*.[2]

1. Statement of the problem.   Let the medium in which the disturbance propagates fill a space $\mathscr{X}$, which for simplicity we take to be n-dimensional Euclidean space.   Thus, every point $x \in \mathscr{X}$ is specified by a set of n real numbers $x^1, \ldots, x^n$.   Choosing a fixed point $x_0 \in \mathscr{X}$, we consider the set of all smooth curves

$$x = x(s) \tag{1}$$

passing through x_0.   The set of vectors tangent to the curve (1) at the point x_0, i.e., the set of vectors

$$x' = \frac{dx}{ds},$$

forms an n-dimensional linear space, which we call the *tangent space* to $\mathscr{X}$ at x_0 and denote by $\mathscr{T}(x_0)$.   Note that the end points of the vectors in any tangent space $\mathscr{T}(x)$ are points of $\mathscr{X}$ itself.[3]

Since the medium is inhomogeneous and anisotropic, the velocity of propagation of disturbances in $\mathscr{X}$ depends on position and direction, i.e., on x and x'.   Let $f(x, x')$ denote the reciprocal of this velocity.   Then, if $x(s)$ and $x(s + ds)$ are two neighboring points lying on some curve $x = x(s)$, the time dt which it takes the disturbance to go from the point $x(s)$ to the point $x(s + ds)$ can be written in the form

$$dt = f\left(x, \frac{dx}{ds}\right) ds,$$

and the time it takes the disturbance to propagate along some infinite path joining the points $x_0 = x(s_0)$ and $x_1 = x(s_1)$ equals

$$\int_{s_0}^{s_1} f\left(x, \frac{dx}{ds}\right) ds. \tag{2}$$

Suppose the point x_0 is "excited," and consider all possible paths joining x_0 and x_1.   Then, because of the "off or on" character of the excitation, the only path which plays any role in the propagation process is the one along which the disturbance propagates in the smallest time, say τ. (Disturbances arriving at x_1 via some other path which is traversed in a time $> \tau$ will arrive

---

[2] See e.g., B. B. Baker and E. T. Copson, *The Mathematical Theory of Huygens' Principle*, Oxford University Press, New York (1939).

[3] In the case considered, the tangent space $\mathscr{T}(x)$ is particularly simple, and in fact, is just an n-dimensional Euclidean space with origin at x.   More generally, $\mathscr{X}$ can be an n-dimensional differentiable manifold, and then the end points of vectors in $\mathscr{T}(x)$ need no longer lie in $\mathscr{X}$.   However, the analysis given below can easily be extended to this case, by exploiting the "local flatness" of $\mathscr{X}$.

at x_1 "too late" to have any further effect on the propagation process, since x_1 will already be found in a state of excitation.) In other words,

$$\tau = \min \int_{s_0}^{s_1} f\left(x, \frac{dx}{ds}\right) ds,$$

where the minimum is taken with respect to all curves $x = x(s)$ joining the points x_0 and x_1. Thus, the propagation of disturbances in the medium obeys the familiar *Fermat principle* (p. 34), i.e., among all paths joining x_0 and x_1, the disturbance always propagates along the path which it traverses in the least time. We shall refer to such paths as the *trajectories* of the disturbance.

Next, we state a physically plausible set of properties for the function $f(x, x')$:

1. The propagation time along any curve is positive, and hence

$$f(x, x') > 0 \quad \text{if} \quad x' \neq 0. \tag{3}$$

2. The propagation time along any curve γ joining x_0 and x_1, given by the integral (2), depends only on γ and not on how γ is parameterized. It follows by the argument given in Chap. 2, Sec. 10 that $f(x, x')$ is positive-homogeneous of degree 1 in x':

$$f(x, \lambda x') = \lambda f(x, x') \quad \text{for every} \quad \lambda > 0. \tag{4}$$

In particular, (4) implies that

$$f(x, x' + \bar{x}') = f(x, x') + f(x, \bar{x}'), \tag{5}$$

if $\bar{x}' = \lambda x'$, where $\lambda > 0$.

3. The time it takes a disturbance to traverse a curve γ connecting x_0 to x_1 is the same as the time it takes a disturbance to traverse γ in the opposite direction from x_1 to x_0, and hence

$$f(x, -x') = f(x, x'). \tag{6}$$

4. If the medium is homogeneous, so that f is a function of direction only, then the disturbance propagates in straight lines (see Prob. 1). In particular, no disturbance emanating from a given point x_0 can arrive at another point x_1 more quickly by taking a path consisting of two straight line segments than by going along the straight line segment joining x_0 and x_1. This implies the *convexity condition*

$$f(x' + \bar{x}') \leqslant f(x') + f(\bar{x}')$$

(see Prob. 2). If f depends on x in a sufficiently smooth way (e.g., if the derivatives $\partial f/\partial x^1, \ldots, \partial f/\partial x^n$ exist), the same argument shows that the convexity condition

$$f(x, x' + \bar{x}') \leqslant f(x, x') + f(x, \bar{x}') \tag{7}$$

holds for sufficiently small x', $\bar{x}'$, but then (7) holds for all x', $\bar{x}'$ because of the homogeneity property (4).

5. Actually, we strengthen the condition (7) somewhat, by requiring that f satisfy the *strict convexity condition*, consisting of (7) plus the stipulation that (5) holds *only* if $\bar{x}' = \lambda x'$, where $\lambda > 0$.

Now suppose we have a disturbance which at time $t = 0$ occupies some region of excitation R in $\mathscr{X}$, and propagates further as time evolves. The boundary of R will be called the *wave front*. Let

$$S(x, t) = 0$$

be the equation of the wave front at the time t. Then our problem can be stated as follows: *Find the equation satisfied by the function $S(x, t)$ describing the wave front, and find the equations of the trajectories of the disturbance.*

2. Introduction of a norm in $\mathscr{T}(x)$. Our next step is to use the function $f(x, x')$ to introduce a norm in the n-dimensional tangent space $\mathscr{T}(x)$. This can be done by defining the norm of the vector $x' = 0$ to be zero and setting

$$\|x'\| = f(x, x') \tag{8}$$

for all vectors $x' \neq 0$ in $\mathscr{T}(x)$. The fact that $\|x'\|$ actually meets all the requirements for a norm (see p. 6) is an immediate consequence of (3), (4), (6) and (7). The set of all vectors in $\mathscr{T}(x)$ such that

$$f(x, x') = \|x'\| = \alpha \tag{9}$$

is called a *sphere of radius α* in $\mathscr{T}(x)$, with center at the point x. The sphere (9) is just the boundary of the closed region of $\mathscr{T}(x)$ [and hence of $\mathscr{X}$] which is excited during the time α by a disturbance originally concentrated at the point x. In this language, our problem can be rephrased as follows: *Suppose a tangent space $\mathscr{T}(x)$, equipped with the norm (8) satisfying the strict convexity condition, is defined at each point x of an n-dimensional space $\mathscr{X}$. Find the equations describing the propagation of disturbances in $\mathscr{X}$, if during the time dt the disturbance originally at x "spreads out and fills" the sphere*

$$f(x, dx) = dt.$$

3. The conjugate space $\tilde{\mathscr{T}}(x)$. Let $\varphi[x']$ be a linear functional (see p. 8), defined on the tangent space $\mathscr{T}(x)$. Then there is a unique vector

$$p = (p_1, \ldots, p_n),$$

such that

$$\varphi[x'] = (p, x')$$

for all $x' \in \mathscr{T}(x)$, where by (p, x') is meant the *scalar product*

$$\sum_{i=1}^{n} p_i x^{1'} + \cdots + p_n x^{n'}$$

(see Prob. 3).[4] Conversely, any scalar product (p, x') obviously defines a linear functional on $\mathscr{T}(x)$. The set of all linear functionals on $\mathscr{T}(x)$, or equivalently the set of all vectors p, is itself an n-dimensional linear space, called the *conjugate space* of $\mathscr{T}(x)$ and denoted by $\tilde{\mathscr{T}}(x)$. We define the *norm* of a vector $p \in \tilde{\mathscr{T}}(x)$ by the formula[5]

$$\|p\| = \sup_{x'} \frac{(p, x')}{\|x'\|}, \tag{10}$$

where the least upper bound is taken over all vectors $x' \neq 0$ in $\mathscr{T}(x)$ [see Prob. 4]. In the present context, we write $H(x, p)$ instead of $\|p\|$, i.e.,

$$H(x, p) = \sup_{x'} \frac{(p, x')}{\|x'\|}. \tag{11}$$

It can be shown that the transition from the function $f(x, x')$ to the function $H(x, p)$ defined by (11) is just the parametric form of the Legendre transformation discussed in Sec. 18.

4. The propagation process. Suppose the wave front at the time t is the surface σ_t, with equation

$$S(x, t) = 0. \tag{12}$$

We now examine in more detail the mechanism governing the evolution of σ_t in time. By hypothesis, each point of σ_t serves as a source of new disturbances, which during the time dt excite the region bounded by the sphere

$$f(x, dx) = dt. \tag{13}$$

Since the function $f(x, x')$ determining the propagation process is assumed to be differentiable and strictly convex (in the sense explained above), there is a unique hyperplane tangent to each point of the sphere (13), and this hyperplane has only one point in common with the sphere. i.e., its point of tangency. If we construct a family of spheres (13), one for each point $x \in \sigma_t$, then the wave front σ_{t+dt} at the time $t + dt$, with equation

$$S(x, t + dt) = 0, \tag{14}$$

is just the envelope E of this family of spheres. In fact, E is the "interface" separating the points of $\mathscr{X}$ which can be reached from σ_t in times $\leqslant dt$ from the points which can only be reached from σ_t in times $> dt$. This construction has two important implications:

---

[4] The reader familiar with tensor analysis will note that here we make a distinction between *contravariant vectors* like x', with components x'^i indexed by superscripts, and *covariant vectors* like p, with components p_i indexed by subscripts. See e.g., G. E. Shilov, *op. cit.*, Sec. 39.

[5] By sup is meant the *least upper bound* or *supremum*.

1. Given a point $x \in \sigma_t$, there is a unique point $x + dx \in \sigma_{t+dt}$ which is excited after the time dt by a disturbance initially at x. In fact, $x + dx$ is the point of σ_{t+dt} lying on the (unique) hyperplane tangent to both (13) and σ_{t+dt}. To see this, we observe that it takes a time $> dt$ for a disturbance starting from x to reach any other point of σ_{t+dt}.[6] Thus, there is a unique direction of propagation defined at each point $x \in \sigma_t$, and it is clear that a disturbance leaving x in this direction will arrive at the surface σ_{t+dt} more quickly than a disturbance leaving x in any other direction, as required by Fermat's principle.

2. Conversely, given a point $x + dx \in \sigma_{t+dt}$, there is a unique point $x \in \sigma_t$, which at the time t was the source of the disturbance reaching $x + dx$ at the time $t + dt$. In fact, x is just the center of the (unique) sphere of radius dt which shares a tangent hyperplane with σ_{t+dt}.

5. The Hamilton-Jacobi equation. As was just shown, every hyperplane tangent to the surface σ_{t+dt} with equation (14) must also be tangent to some sphere of radius dt whose center lies on the surface σ_t with equation (12). This fact can be used to derive a differential equation satisfied by the function $S(x, t)$. First, we observe that every hyperplane in the tangent space $\mathscr{T}(x)$ can be written in the form

$$\sum_{i=1}^{n} p_i x^{i\prime} = \text{const},$$

where $p = (p_1, \ldots, p_n)$ is a vector in the conjugate space $\tilde{\mathscr{T}}(x)$. Let $x + dx$ be an arbitrary point of σ_{t+dt}, whose "source" is the point $x \in \sigma_t$. Then the hyperplane in $\mathscr{T}(x)$ tangent to σ_{t+dt} at $x + dx$ has the equation

$$\sum_{i=1}^{n} \frac{\partial S}{\partial x^i} dx^i = c, \tag{15}$$

where c is a constant. If the hyperplane (15) is also tangent to the sphere (13), as required, then c equals the norm of the vector

$$\nabla S = \left(\frac{\partial S}{\partial x^1}, \ldots, \frac{\partial S}{\partial x^n} \right),$$

multiplied by the radius of the sphere, i.e.,

$$c = H(x, \nabla S) \, dt.$$

Therefore, (15) becomes

$$\sum_{i=1}^{n} \frac{\partial S}{\partial x^i} dx^i = H(x, \nabla S) \, dt. \tag{16}$$

---

[6] Physically, this means that if the surface σ_t is changed only in a small neighborhood of the point x, the surface σ_{t+dt} is also changed only in a small neighborhood of $x + dx$.

But

$$\sum_{i=1}^{n} \frac{\partial S}{\partial x^i} \, dx^i + \frac{\partial S}{\partial t} \, dt = 0, \tag{17}$$

because of the meaning of x and $x + dx$. Comparing (16) and (17), we finally obtain

$$\frac{\partial S}{\partial t} + H(x, \nabla S) = 0. \tag{18}$$

This equation describes the way the wave front evolves in time, and is just the familiar *Hamilton-Jacobi equation*, already considered in Sec. 23.

We now show the relation between the trajectories of the disturbance and the general solution of (18). It will be recalled that as a wave front evolves in time, each of its points goes into a succession of uniquely defined points lying on neighboring wave fronts, thereby "sweeping out" a trajectory γ which automatically minimizes the functional (2). Thus, if we specify a one-parameter family of wave fronts

$$S(x, t) = 0, \tag{19}$$

where the parameter is the time t, every point x_0 on some "initial" surface $S(x, t_0)$ generates a trajectory. Choosing the point x_0 arbitrarily, we find that the one-parameter family of surfaces (19) determines an $(n - 1)$-parameter family of trajectories, such that one and only one trajectory of the family passes through each point $x \in \mathscr{X}$. More generally, let

$$S(x, t, \alpha_1, \ldots, \alpha_n)$$

be a complete integral of the Hamilton-Jacobi equation depending on n parameters $\alpha_1, \ldots, \alpha_n$. This complete integral determines an $(n + 1)$-parameter family of surfaces[7]

$$S(x, t, \alpha_1, \ldots, \alpha_n) = 0, \tag{20}$$

which in turn determines a $(2n - 1)$-parameter family of trajectories. Then the fact that the trajectories of the disturbances are the extremals of the functional (2) leads to a geometric interpretation of Jacobi's theorem (p. 91), concerning the construction of a general solution of the system of Euler equations of a functional from a complete integral of the corresponding Hamilton-Jacobi equation.[8]

---

[7] Since $S(x, t + t_0, \alpha_1, \ldots, \alpha_n) = 0$ is also an integral surface of the Hamilton-Jacobi equation for arbitrary t_0, the family of surfaces (20) actually depends on $n + 1$ parameters.

[8] It should be noted that we are considering a parametric problem, so that there is dependence between the Euler equations (see Sec. 10 and Remark 4 of Sec. 37). As a result, the general solution of the $2n$ equations obtained here contains only $2n - 1$ arbitrary constants.

6. The canonical equations. To derive the differential equations satisfied by the trajectories of the disturbance, we might use Fermat's principle, minimizing the functional (2) and solving the corresponding Euler equations. However, we prefer to use our geometric model of the propagation process. If we introduce the time t as the parameter along each trajectory, it follows from

$$f(x, dx) = dt$$

and the homogeneity of $f(x, dx)$ in the argument dx that

$$f\left(x, \frac{dx}{dt}\right) = 1, \tag{21}$$

i.e., the norm of the vector dx/dt is identically equal to 1. Using (16), we find that at each point x, the vector dx/dt (tangent to the trajectory along which the disturbance propagates) is related to the covariant vector p (determining the hyperplane tangent to the wave front) by the formula

$$\sum_{i=1}^{n} p_i \frac{dx^i}{dt} = H(x, p).$$

According to (21) and the definition (11) of the norm of vectors in $\tilde{\mathcal{T}}(x)$, we see that

$$\sum_{i=1}^{n} \frac{dx^i}{dt} \leqslant H(x, p)$$

if p is any other vector in $\tilde{\mathcal{T}}(x)$. Thus, the expression

$$\sum_{i=1}^{n} \frac{dx^i}{dt} - H(x, p),$$

regarded as a function of p, achieves its maximum when p is the vector determining the hyperplane tangent to the wave front. Therefore, along the trajectories, the conditions

$$\frac{\partial}{\partial p_i}\left[\sum_{i=1}^{n} p_i \frac{dx^i}{dt} - H(x, p)\right] = 0 \qquad (i = 1, \ldots, n)$$

must hold, i.e.,

$$\frac{dx^i}{dt} = \frac{\partial H(x, p)}{\partial p_i} \qquad (i = 1, \ldots, n). \tag{22}$$

We have just obtained a system of n ordinary differential equations of the first order satisfied by the trajectories. Since these equations involve $2n$ unknown functions $x^1, \ldots, x^n$ and $p_1, \ldots, p_n$, we still need n more equations to completely describe the trajectories. To find the missing equations, we use the fact that the surfaces representing the wave fronts at different times are not arbitrary, but satisfy the Hamilton-Jacobi equation (18), while the values p_i at each point of a trajectory are the components $\partial S/\partial x^i$ determining the hyperplane tangent to the wave front.   In other words,

$$p_i = p_i(t) = \frac{\partial}{\partial x^i} S[x^1(t), \ldots, x^n(t), t]$$

along each trajectory, and hence

$$\frac{dp_i}{dt} = \frac{d}{dt} \frac{\partial S}{\partial x^i} = \frac{\partial}{\partial t} \frac{\partial S}{\partial x^i} + \sum_{k=1}^{n} \frac{\partial^2 S}{\partial x^k \partial x^i} \frac{dx^k}{dt}. \tag{23}$$

We now introduce the following notation: If the function $H(x, p)$, where $p_i = \partial S/\partial x_i$, is regarded as a function of $x^1, \ldots, x^n$ and t, we indicate its partial derivative with respect to x^i by

$$\frac{\partial H}{\partial x^i}\bigg|_{t=\text{const}},$$

whereas if $H(x, p)$ is regarded as a function of the $2n$ variables $x^1, \ldots, x^n$ and $p_1, \ldots, p_n$, we indicate its partial derivative with respect to x^i by

$$\frac{\partial H}{\partial x^i}\bigg|_{p=\text{const}}.$$

Then, using the Hamilton-Jacobi equation (18), we can write (23) in the form

$$\frac{dp_i}{dt} = -\frac{\partial H}{\partial x^i}\bigg|_{t=\text{const}} + \sum_{k=1}^{n} \frac{\partial^2 S}{\partial x^k \partial x^i} \frac{dx^k}{dt}. \tag{24}$$

Along the trajectories, we have

$$\frac{\partial H}{\partial x^i}\bigg|_{t=\text{const}} = \frac{\partial H}{\partial x^i}\bigg|_{p=\text{const}} + \sum_{k=1}^{n} \frac{\partial H}{\partial p_k}\bigg|_{x=\text{const}} \frac{\partial p_k}{\partial x^i}, \tag{25}$$

and

$$p_k = \frac{\partial S}{\partial x^k}, \qquad \frac{dx^k}{dt} = \frac{\partial H}{\partial p_k}. \tag{26}$$

Substituting (25) and (26) into (24), we obtain n differential equations

$$\frac{dp_i}{dt} = -\frac{\partial H}{\partial x^i}\bigg|_{p=\text{const}} \qquad (i = 1, \ldots, n).$$

Combining these equations with (22), we obtain a system of $2n$ differential equations

$$\frac{dx^i}{dt} = \frac{\partial H(x, p)}{\partial p_i},$$

$$\frac{dp_i}{dt} = -\frac{\partial H(x, p)}{\partial x^i}, \tag{27}$$

where $i = 1, \ldots, n$.   The integral curves of (27) are the trajectories along which the disturbance propagates, i.e., the extremals of the functional (2).

The system (27) is of course the *canonical system* of Euler equations for the variational problem associated with (2) [cf. Sec. 16], and represents the so-called *characteristic system* associated with the Hamilton-Jacobi equation (18) [cf. p. 90].

PROBLEMS

1. Prove that if $f(x, x')$ depends on direction only, then the disturbance propagates through the medium along straight lines.

2. Prove that if $f(x, x') \equiv f(x')$ is independent of x, then $f(x')$ is precisely the time required to traverse the vector x'.

3. Prove that every linear functional $\varphi[x]$ defined on an n-dimensional Euclidean space of points $x = (x^1, \ldots, x^n)$ is of the form

$$\varphi[x] = p_1 x^1 + \cdots + p_n x^n,$$

where $p = (p_1, \ldots, p_n)$ is uniquely determined by φ.

4. Verify that formula (10) actually defines a norm for the elements p of the conjugate space $\tilde{\mathscr{T}}(x)$.

5. Why is the strict convexity condition (p. 211) needed in constructing wave fronts for the disturbance?

VARIATIONAL METHODS
IN PROBLEMS OF
OPTIMAL CONTROL

In this appendix, we sketch some results obtained by L. S. Pontryagin and his students, in their investigations of the theory of *optimal control processes*.[1] The connection between this subject and classical variational theory will also be discussed.

1. Statement of the problem. In many cases, finding the optimal "operating regime" for a physical system (with a suitable optimality criterion) leads to the following mathematical problem: Suppose the state of the physical system is characterized by n real numbers $x^1, \ldots, x^n$, forming a vector $x = (x^1, \ldots, x^n)$ in the n-dimensional "phase space" $\mathscr{X}$ of the system, and suppose the state varies with time in the way described by the system of differential equations

$$\frac{dx^i}{dt} = f^i(x^1, \ldots, x^n, u^1, \ldots, u^k) \qquad (i = 1, \ldots, n). \tag{1}$$

Here, the k real numbers $u^1, \ldots, u^k$ form a vector $u = (u^1, \ldots, u^k)$ belonging to some fixed "control region" Ω, which we take to be a subset of

---

[1] See L. S. Pontryagin, *Optimal control processes*, Usp. Mat. Nauk, **14**, no. 1, 3 (1959); V. G. Boltyanski, R. V. Gamkrelidze and L. S. Pontryagin, *The theory of optimal processes, I, The maximum principle*, Izv. Akad. Nauk SSSR, Ser. Mat., **24**, 3 (1960); L. S. Pontryagin, V. G. Boltyanski, R. V. Gamkrelidze and E. F. Mishchenko, *The Mathematical Theory of Optimal Processes*, translated and edited by K. N. Trirogoff and L. W. Neustadt, Interscience Publishers, New York (1962). The more general case where Ω is a topological space is considered in the first two references.

k-dimensional Euclidean space, and the $f^i(x, u)$ are n continuous functions defined for all $x \in \mathscr{X}$ and all $u \in \Omega$.

Now suppose we specify a vector function $u(t)$, $t_0 \leqslant t \leqslant t_1$, called the *control function*, with values in Ω. Then, substituting $u = u(t)$ in (1), we obtain the system of differential equations

$$\frac{dx^i}{dt} = f^i[x^1, \ldots, x^n, u^1(t), \ldots, u^k(t)] \qquad (i = 1, \ldots, n). \qquad (2)$$

For every initial value $x_0 = x(t_0)$, this system has a definite solution, called a *trajectory*. The aggregate

$$U = \{u(t), t_0, t_1, x_0\}, \qquad (3)$$

consisting of a control function $u(t)$, an interval $[t_0, t_1]$ and an initial value $x_0 = x(t_0)$, will be called a *control process*. Thus, to every control process, there corresponds a trajectory, i.e., a solution of (2).

Next, let

$$f^0(x^1, \ldots, x^n, u^1, \ldots, u^k)$$

be a function which is defined, together with its partial derivatives

$$\frac{\partial f^0}{\partial x^i} \qquad (i = 1, \ldots, n),$$

for all $x \in \mathscr{X}$ and $u \in \Omega$. To every control process U, we assign the number

$$J[U] = \int_{t_0}^{t_1} f^0(x, u)\, dt, \qquad (4)$$

i.e., $J[U]$ is a functional defined on the set of control processes. Then, the control process (3) is said to be *optimal* if the inequality

$$J[U] \leqslant J[U^*]$$

holds for any other control process U^* carrying the given point x_0 into the point x_1, i.e., such that the corresponding trajectory $x^*(t)$ satisfies the condition $x^*(t_1^*) = x_1$. By the *optimal trajectory*, we mean the trajectory corresponding to the optimal control process. Our aim is to find necessary conditions characterizing optimal control processes and optimal trajectories.

It should be pointed out that in calling a control process *optimal*, it is assumed that some class of *admissible* control processes has been specified in advance. Here, we assume that the components $u^1(t), \ldots, u^k(t)$ of any admissible control process take values in Ω, and are bounded and piecewise continuous (with left-hand and right-hand limits at every point of discontinuity).

An important special case of the problem of optimal control is the situation where the functional (4) reduces to the integral

$$\int_{t_0}^{t_1} dt,$$

representing the time it takes to go from the point x_0 to the point x_1. In this case, optimality means taking the least time to go from x_0 to x_1.

2. Relation to the calculus of variations. The problem of optimal control is intimately related to certain traditional problems of the calculus of variations. In fact, the integral

$$\int_{t_0}^{t_1} f^0(x, u)\, dt$$

can be regarded as a functional depending on $n + k$ functions $x^1, \ldots, x^n$, $u^1, \ldots, u^k$, i.e., as a functional defined on some class of curves in $n + k + 1$ dimensions. Since the functions $x^1, \ldots, x^n, u^1, \ldots, u^k$ are connected by the equations (1), we are dealing with the problem of finding a minimum subject to nonholonomic constraints (see p. 48). Since the boundary conditions are equivalent to the requirement that the desired optimal trajectory $x(t)$ begin at the point x_0 and end at the point x_1, the end points of the admissible curves in our $(n + k + 1)$-dimensional space have to lie on two $(k + 1)$-dimensional hyperplanes, determined by giving the coordinates $x^1, \ldots, x^n$ the fixed values $x_0^1, \ldots, x_0^n$ and $x_1^1, \ldots, x_1^n$.

Thus, we see that the problem of optimal control is a variant of the problem of finding a minimum subject to subsidiary conditions. The problem of optimal control has the special feature that we specify in advance a definite class of admissible control processes, where the functions $u^1(t), \ldots, u^k(t)$ are required to take values in a given fixed region Ω, but in general are not required to be continuous.

We can easily show that the simplest n-dimensional variational problem, where the integrand does not depend on t explicitly,[2] is a special case of the problem of optimal control. To this end, suppose that among the curves passing through two fixed points

$$(x_0^1, \ldots, x_0^n), \qquad (x_1^1, \ldots, x_1^n),$$

it is required to find the curve for which the functional

$$\int_{t_0}^{t_1} f^0\left(x^1, \ldots, x^n, \frac{dx^1}{dt}, \ldots, \frac{dx^n}{dt}\right) dt \tag{5}$$

has a minimum. To paraphrase this problem as a problem of optimal control, we need only write (5) in the form

$$\int_{t_0}^{t_1} f^0(x^1, \ldots, x^n, u^1, \ldots, u^k)\, dt,$$

and take the system (1) to be simply

$$\frac{dx^i}{dt} = u^i \qquad (i = 1, \ldots, n).$$

---

[2] This condition is not really a restriction, since any functional can be transformed into this form, e.g., by going over to the parametric form of the problem.

3. Necessary conditions for optimality.   To find necessary conditions for a given control process and the corresponding trajectory to be optimal, we supplement the system of equations

$$\frac{dx^i}{dt} = f^i(x, u) \qquad (i = 1, \ldots, n)$$

with the extra equation

$$\frac{dx^0}{dt} = f^0(x, u),$$

where $f^0(x, u)$ is the integrand of the functional (4) which is to be minimized. At the same time, we supplement the initial conditions

$$x^i(t_0) = x_0^i \qquad (i = 1, \ldots, n) \tag{6}$$

with the extra condition

$$x^0(t_0) = 0. \tag{7}$$

For convenience, we introduce the $(n + 1)$-dimensional vector function

$$\mathbf{x}(t) = (x^0(t), x(t)) = (x^0(t), x^1(t), \ldots, x^n(t)).$$

It is clear that if U is an admissible control process and if $\mathbf{x} = \mathbf{x}(t)$ is the solution of the system[3]

$$\frac{dx^i}{dt} = f^i(x, u) \qquad (i = 0, 1, \ldots, n), \tag{8}$$

corresponding to U and the initial conditions (6) and (7), then

$$J[U] = \int_{t_0}^{t_1} f^0(x, u)\, dt = x^0(t_1).$$

Thus, the problem of optimal control can be stated as follows: Find the admissible control process U for which the solution $\mathbf{x}(t)$ of the system (8), satisfying the initial conditions (6) and (7), has the smallest possible value of $x^0(t_1)$.

Next, in addition to the variables $x^0, x^1, \ldots, x^n$, we introduce new variables $\psi_0, \psi_1, \ldots, \psi_n$ satisfying the following system of differential equations, known as the *conjugate*[4] of the system (8):

$$\frac{d\psi_i}{dt} = -\sum_{\alpha=0}^{n} \frac{\partial f^\alpha(x, u)}{\partial x^i}\, \psi_\alpha \qquad (i = 0, 1, \ldots, n). \tag{9}$$

---

[3] Note that the functions f^α, and hence the functions Π and H defined below, do not involve $x^0(t)$.

[4] This system has the following geometric interpretation: In the space of vectors $(\psi_0, \psi_1, \ldots, \psi_n)$ conjugate to the space of vectors $(x^0, x^1, \ldots, x^n)$ [see p. 211], consider the hyperplane

$$\sum_{\alpha=0}^{n} \psi_\alpha^0 x_\alpha = c = \text{const}$$

passing through the initial point $(0, x_0^1, \ldots, x_0^n)$.   Then the system (9) describes the "transport" of this hyperplane along the trajectories corresponding to solutions of the system (8).   In other words, if the ψ_i satisfy (9) and the x^i satisfy (9) for $t_0 \leqslant t \leqslant t_1$, then

$$\sum_{\alpha=0}^{n} \psi_\alpha x^\alpha = c \qquad (t_0 \leqslant t \leqslant t_1).$$

For more details, see the second of the references cited on p. 218.

Let

$$\psi(t) = (\psi_0(t), \psi_1(t), \ldots, \psi_n(t)),$$

and consider the following function of the variables $x^1, \ldots, x^n, \psi_0, \psi_1, \ldots, \psi_n, u_1, \ldots, u_k$:

$$\Pi(\psi, x, u) = \sum_{\alpha=0}^{n} \psi_\alpha f^\alpha(x, u). \tag{10}$$

In terms of Π, we can write the equations (8) and (9) in the form

$$\frac{dx^i}{dt} = \frac{\partial \Pi}{\partial \psi_i},$$
$$\frac{d\psi_i}{dt} = -\frac{\partial \Pi}{\partial x^i}, \tag{11}$$

where $i = 0, 1, \ldots, n$. The equations (11) remind us of the canonical system of Euler equations [see formula (11), p. 70]. However, they have a different meaning, since the canonical equations form a closed system, in which the number of equations equals the number of unknown functions, whereas (11) involves not only $\mathbf{x}$ and ψ but also the unknown function u, and hence (10) becomes a closed system only when u is specified. In fact, in order to write equations for the optimal control problem resembling the canonical equations, we would have to use the function

$$\mathscr{H}(\psi, x) = \sup_{u\in\Omega} \Pi(\psi, x, u), \tag{12}$$

instead of the function $\Pi(\psi, x, u)$.[5]

4. The maximum principle. We can now state the following theorem, whose proof can be found in the references cited on p. 218:

THEOREM (*The maximum principle*). *Let* $U = \{u(t), t_0, t_1, x_0\}$ *be an admissible control process, and let* $\mathbf{x}(t)$ *be the corresponding integral curve of the system* (8) *passing through the point* $(0, x_0^1, \ldots, x_0^n)$ *for* $t = 0$, *and satisfying the conditions*

$$x^1(t_1) = x_1^1, \ldots, x^n(t_1) = x_1^n$$

for $t = t_1$. *Then if the control process* U *is optimal, there exists a continuous vector function* $\psi(t) = (\psi_0(t), \psi_1(t), \ldots, \psi_n(t))$ *such that*

1. *The function* $\psi(t)$ *satisfies the system* (9) *for* $x = x(t), u = u(t)$;

---

[5] The transition from Π to $\mathscr{H}$ is analogous to the Legendre transformation, considered in Sec. 18.

2. *For all t in $[t_0, t_1]$, the function (10) achieves its maximum for $u = u(t)$, i.e.,*

$$\Pi[\psi(t), x(t), u(t)] = \mathcal{H}[\psi(t), x(t)], \tag{13}$$

where the function $\mathcal{H}$ is defined by (12);

3. *The relations*

$$\psi_0(t_1) \leqslant 0, \quad \mathcal{H}[\psi(t_1), u(t_1)] = 0 \tag{14}$$

hold at the time t_1. Actually, if $\psi(t)$, $\mathbf{x}(t)$ and $u(t)$ satisfy the system (8), (9) and the condition (13), the functions $\psi_0(t)$ and $\mathcal{H}[\psi(t), x(t)]$ turn out to be constants, and hence in (14) we can replace t_1 by any value of t in $[t_0, t_1]$.

Remark 1. The maximum principle can often be used as a prescription for constructing the optimal trajectory, in the following way: For every fixed ψ and x, we find the value of u for which the expression

$$\sum_{\alpha=0}^{n} \psi_\alpha f^\alpha(x, u)$$

takes its maximum. If this determines u as a single-valued function

$$u = u(\psi, x) \tag{15}$$

of ψ and x, then, substituting (15) into the equations (8) and (9), we obtain a closed system of $2(n + 1)$ equations involving $2(n + 1)$ unknown functions. These are just the equations which have to be satisfied by the optimal trajectory.

Remark 2. For the simple n-dimensional variational problem discussed on p. 220, the system (8), (9), or the equivalent system (11), together with the maximum principle, reduces to the usual system of Euler equations. To see this, consider the functional

$$\int_{t_0}^{t_1} f^0(x^1, \ldots, x^n, u^1, \ldots, u^n) \, dt \tag{16}$$

[cf. (5)], where

$$u^i = \frac{dx^i}{dt} \qquad (i = 1, \ldots, n). \tag{17}$$

In this case, the function (10) is

$$\Pi(\psi, x, u) = \psi_0 f^0(x, u) + \sum_{\alpha=1}^{n} \psi_\alpha u^\alpha, \tag{18}$$

and the system (11) becomes

$$\frac{dx^0}{dt} = f^0(x, u), \qquad \frac{dx^i}{dt} = u^i,$$

$$\frac{d\psi_0}{dt} = 0, \qquad \frac{d\psi_i}{dt} = -\psi_0 \frac{\partial f^0(x, u)}{\partial x^i},$$

where $i = 1, \ldots, n$. Maximizing $\Pi(\psi, x, u)$, we find that

$$\frac{\partial \Pi}{\partial u^i} = \psi_0 \frac{\partial f^0(x, u)}{\partial u^i} + \psi_i = 0,$$

i.e.,

$$\psi_i = -\psi_0 \frac{\partial f^0(x, u)}{\partial u^i} \qquad (i = 1, \ldots, n).$$

Since $d\psi_0/dt = 0$, we have $\psi_0 = \text{const}$, and hence

$$\frac{d}{dt} \left[\frac{\partial f^0(x, u)}{\partial u^i} \right] = \frac{\partial f^0(x, u)}{\partial x^i},$$

$$\frac{dx^i}{dt} = u^i.$$

This is just the system of Euler equations corresponding to the functional (16), reduced to a system of first-order differential equations by introducing the derivatives $dx^i/dt = u^i$ as new functions (cf. p. 68).

Remark 3. In Appendix I, we have already encountered the fact that every propagation process can be described in two ways, either in terms of the trajectories along which the disturbance propagates (the "rays" in optics), or in terms of the motion of the wave front. The first approach leads to the canonical Euler equations (or, as in the example just considered, to the usual form of the Euler equations), i.e., a system of ordinary differential equations. The second approach leads to the Hamilton-Jacobi equation, i.e., a partial differential equation. Our maximum principle involves the study of trajectories, and in this sense is analogous to the method of canonical equations. The "wave front approach" to problems of optimal control has been developed by R. Bellman.[6]

5. Relation to Weierstrass' necessary condition. We again consider the simple functional (16), (17), where the function $\Pi(\psi, x, u)$ is given by (18). Using (17), we can also write the functional (16) in the form

$$\int_{t_0}^{t_1} f^0(x^1, \ldots, x^n, x^{1\prime}, \ldots, x^{n\prime}) \, dt. \tag{19}$$

The Weierstrass E-function for such a functional is[7]

$$E(x, x', z) = f^0(x, z) - f^0(x, x') - \sum_{i=1}^{n} (z - x_i')f_{x^{i\prime}}^0(x, x'). \tag{20}$$

---

[6] See the relevant references cited in the Bibliography, p. 227.

[7] See p. 146. Note that E is a function of three rather than four arguments, since (19) is independent of t.

Using (18) and (20), we find that

$$\Pi(\psi, x, z) - \Pi(\psi, x, x') - \sum_{i=1}^{n} (z_i - x^{i'}) \frac{\partial}{\partial u^i} \Pi(\psi, x, x')$$

$$= \psi_0 f^0(x, z) - \psi_0 f^0(x, x') + \sum_{i=1}^{n} \psi_i(z_i - x^{i'}) - \sum_{i=1}^{n} (z_i - x^{i'})(\psi_0 f^0_{x^{i'}} + \psi_i)$$

$$= \psi_0 f^0(x, z) - \psi_0 f^0(x, x') - \sum_{i=1}^{n} (z_i - x^{i'})\psi_0 f^0_{x^{i'}} = \psi_0 E(x, x', z). \quad (21)$$

If the function Π achieves its maximum for values of $u = x'$ which are interior points of the region Ω, then

$$\frac{\partial \Pi}{\partial u^i} = 0$$

at these points. Then, since $\psi_0 \leqslant 0$, it follows from (21) that the condition (13) is equivalent to the condition

$$E(x, x', z) \geqslant 0. \quad (22)$$

This is *Weierstrass' necessary condition*, with which we are already familiar (see p. 149). Thus, the maximum principle leads to another, independent derivation of (22). It can be shown that the formula

$$\psi_0 E = \Pi(\psi, x, z) - \Pi(\psi, x, x') - \sum_{i=1}^{n} (z_i - x^{i'}) \frac{\partial}{\partial u^i} \Pi(\psi, x, x')$$

remains true for variational problems subject to constraints, i.e., for more general problems of optimal control.

We have just proved the equivalence of the maximum principle and Weierstrass' necessary condition (22) in the case where the set Ω of admissible values of the control function $u(t)$ is open, i.e., where every point of Ω is an interior point. In the case where the optimal control process involves values of $u(t)$ lying on the boundary of the region Ω, the condition (22) is in general no longer valid. However, it can be shown that in such cases, the maximum principle continues to apply.

PROBLEMS

1. State the maximum principle (p. 222) for the problem of "fastest motion" or "time optimal problem," where the functional (4) reduces to simply

$$J[U] = \int_{t_0}^{t_1} dt.$$

Ans. In this case, we write

$$P(\psi, x, u) = \sum_{\alpha=1}^{n} \psi_\alpha f^\alpha(x, u)$$

instead of (10), and in the system (11), i need only range from 1 to n. The function $\mathscr{H}$ in the maximum principle is now replaced by

$$H(\psi, x) = \sup_{u \varepsilon \Omega} P(\psi, x, u) = \mathscr{H}(\psi, x) - \psi_0.$$

Finally, the relations (14) are replaced by

$$H[\psi(t_1), x(t_1)] = -\psi_0 \geqslant 0,$$

which actually holds for any t in $[t_0, t_1]$.

2. Consider the differential equation

$$\frac{d^2x}{dt^2} = u, \tag{a}$$

where the control function u obeys the condition $|u| \leqslant 1$. Introducing the "phase coordinates" x^1 and x^2, we can write (a) as a system

$$\frac{dx^1}{dt} = x^2, \qquad \frac{dx^2}{dt} = u. \tag{b}$$

What trajectory corresponds to the fastest motion from a given initial point x_0 to the final point $x_1 = (0, 0)$?

Hint. The auxiliary variables ψ_1 and ψ_2 obey the equations

$$\frac{d\psi_1}{dt} = 0, \qquad \frac{d\psi_2}{dt} = -\psi_1.$$

By the maximum principle (modified in accordance with Prob. 1),

$$u(t) = \operatorname{sgn} \psi_2(t) = \operatorname{sgn}(c_2 - c_1 t),$$

where c_1 and c_2 are constants, $\operatorname{sgn} x \equiv x/|x|$ and $u(t)$ can only change sign once. Integrate the system (b) for $u = \pm 1$, and draw the corresponding families of parabolas in the (x^1, x^2) plane, analyzing the various possibilities (corresponding to different initial positions x_0).

3. Study the same "time-optimal problem" for the equation

$$\frac{d^2x}{dt^2} + x = u, \qquad |u| \leqslant 1.$$

Hint. The appropriate system is now

$$\frac{dx^1}{dt} = x^2, \qquad \frac{dx^2}{dt} = -x^1 + u.$$

4. Study the same "time-optimal problem" for the system

$$\frac{dx^1}{dt} = x^2 + u^1, \qquad \frac{dx^2}{dt} = -x^1 + u^2,$$

where there are *two* control functions u^1, u^2 obeying the conditions $|u^1| \leqslant 1$, $|u^2| \leqslant 1$.

Comment. For a detailed discussion of Probs. 2–4, see Chap. 1, Sec 5 of the book cited on p. 218.

5. Verify the relations (14) for the simple variational problem (16) discussed in Remark 2, p. 223.

Hint. Use Euler's theorem on positive-homogeneous functions (Chap. 2, Prob. 6).

BIBLIOGRAPHY[1]

Akhiezer, N. I., *The Calculus of Variations*, translated by A. H. Frink, Blaisdell Publishing Co., New York (1962).

Bellman, R., *Dynamic Programming*, Princeton University Press, Princeton, N. J. (1957).

Bellman, R., *Adaptive Control Processes: A Guided Tour*, Princeton University Press, Princeton, N. J. (1961).

Bellman, R. and S. E. Dreyfus, *Applied Dynamic Programming*, Princeton University Press, Princeton, N. J. (1962).

Bliss, G. A., *Calculus of Variations*, Open Court Publishing Co., Chicago (1925).

Bliss, G. A., *Lectures on the Calculus of Variations*, University of Chicago Press, Chicago (1946).

Bolza, O., *Lectures on the Calculus of Variations*, reprinted by G. E. Stechert and Co., New York (1931).

Courant, R. and D. Hilbert, *Methods of Mathematical Physics*, *Vol. I*, Interscience Publishers, Inc., New York (1953), Chaps. 4 and 6.

Elsgolc, L. E., *Calculus of Variations*, translated from the Russian, Addison-Wesley Publishing Co., Reading, Mass. (1962).

Forsyth, A. R., *Calculus of Variations*, reprinted by Dover Publications, Inc., New York (1960).

Fox, C., *An Introduction to the Calculus of Variations*, Oxford University Press, New York (1950).

Gould, S. H., *Variational Methods for Eigenvalue Problems*, University of Toronto Press, Toronto (1957).

Lanczos, C., *The Variational Principles of Mechanics*, University of Toronto Press, Toronto (1949).

Morrey, C. B. Jr., *Multiple Integral Problems in the Calculus of Variations and Related Topics*, University of California Press, Berkeley and Los Angeles (1943).

Morse, M., *The Calculus of Variations in the Large*, American Mathematical Society, Providence, R. I. (1934).

Murnaghan, F. D., *The Calculus of Variations*, Spartan Books, Washington, D.C. (1962).

Pars, L. A., *Calculus of Variations*, John Wiley and Sons, Inc., New York (1963).

Weinstock, R., *Calculus of Variations, with Applications to Physics and Engineering*, McGraw-Hill Book Co., Inc., New York (1952).

---

[1] See also books cited on pp. 205 and 218.

INDEX

A CATALOG OF SELECTED
DOVER BOOKS
IN SCIENCE AND MATHEMATICS

A CATALOG OF SELECTED

DOVER BOOKS

IN SCIENCE AND MATHEMATICS

Astronomy

BURNHAM'S CELESTIAL HANDBOOK, Robert Burnham, Jr. Thorough guide to the stars beyond our solar system. Exhaustive treatment. Alphabetical by constellation: Andromeda to Cetus in Vol. 1; Chamaeleon to Orion in Vol. 2; and Pavo to Vulpecula in Vol. 3. Hundreds of illustrations. Index in Vol. 3. 2,000pp. 6⅛ x 9¼.
23567-X, 23568-8, 23673-0 Three-vol. set

THE EXTRATERRESTRIAL LIFE DEBATE, 1750–1900, Michael J. Crowe. First detailed, scholarly study in English of the many ideas that developed from 1750 to 1900 regarding the existence of intelligent extraterrestrial life. Examines ideas of Kant, Herschel, Voltaire, Percival Lowell, many other scientists and thinkers. 16 illustrations. 704pp. 5⅜ x 8½. 40675-X

A HISTORY OF ASTRONOMY, A. Pannekoek. Well-balanced, carefully reasoned study covers such topics as Ptolemaic theory, work of Copernicus, Kepler, Newton, Eddington's work on stars, much more. Illustrated. References. 521pp. 5⅜ x 8½.
65994-1

AMATEUR ASTRONOMER'S HANDBOOK, J. B. Sidgwick. Timeless, comprehensive coverage of telescopes, mirrors, lenses, mountings, telescope drives, micrometers, spectroscopes, more. 189 illustrations. 576pp. 5⅝ x 8¼. (Available in U.S. only.)
24034-7

STARS AND RELATIVITY, Ya. B. Zel'dovich and I. D. Novikov. Vol. 1 of *Relativistic Astrophysics* by famed Russian scientists. General relativity, properties of matter under astrophysical conditions, stars, and stellar systems. Deep physical insights, clear presentation. 1971 edition. References. 544pp. 5⅝ x 8¼. 69424-0

Chemistry

CHEMICAL MAGIC, Leonard A. Ford. Second Edition, Revised by E. Winston Grundmeier. Over 100 unusual stunts demonstrating cold fire, dust explosions, much more. Text explains scientific principles and stresses safety precautions. 128pp. 5⅜ x 8½. 67628-5

THE DEVELOPMENT OF MODERN CHEMISTRY, Aaron J. Ihde. Authoritative history of chemistry from ancient Greek theory to 20th-century innovation. Covers major chemists and their discoveries. 209 illustrations. 14 tables. Bibliographies. Indices. Appendices. 851pp. 5⅜ x 8½. 64235-6

CATALYSIS IN CHEMISTRY AND ENZYMOLOGY, William P. Jencks. Exceptionally clear coverage of mechanisms for catalysis, forces in aqueous solution, carbonyl- and acyl-group reactions, practical kinetics, more. 864pp. 5⅜ x 8½.
65460-5

THE HISTORICAL BACKGROUND OF CHEMISTRY, Henry M. Leicester. Evolution of ideas, not individual biography. Concentrates on formulation of a coherent set of chemical laws. 260pp. 5⅜ x 8½. 61053-5

A SHORT HISTORY OF CHEMISTRY, J. R. Partington. Classic exposition explores origins of chemistry, alchemy, early medical chemistry, nature of atmosphere, theory of valency, laws and structure of atomic theory, much more. 428pp. 5⅜ x 8½. (Available in U.S. only.) 65977-1

GENERAL CHEMISTRY, Linus Pauling. Revised 3rd edition of classic first-year text by Nobel laureate. Atomic and molecular structure, quantum mechanics, statistical mechanics, thermodynamics correlated with descriptive chemistry. Problems. 992pp. 5⅜ x 8½. 65622-5

Engineering

DE RE METALLICA, Georgius Agricola. The famous Hoover translation of greatest treatise on technological chemistry, engineering, geology, mining of early modern times (1556). All 289 original woodcuts. 638pp. 6¾ x 11. 60006-8

FUNDAMENTALS OF ASTRODYNAMICS, Roger Bate et al. Modern approach developed by U.S. Air Force Academy. Designed as a first course. Problems, exercises. Numerous illustrations. 455pp. 5⅜ x 8½. 60061-0

DYNAMICS OF FLUIDS IN POROUS MEDIA, Jacob Bear. For advanced students of ground water hydrology, soil mechanics and physics, drainage and irrigation engineering and more. 335 illustrations. Exercises, with answers. 784pp. 6⅛ x 9¼. 65675-6

ANALYTICAL MECHANICS OF GEARS, Earle Buckingham. Indispensable reference for modern gear manufacture covers conjugate gear-tooth action, gear-tooth profiles of various gears, many other topics. 263 figures. 102 tables. 546pp. 5⅜ x 8½. 65712-4

MECHANICS, J. P. Den Hartog. A classic introductory text or refresher. Hundreds of applications and design problems illuminate fundamentals of trusses, loaded beams and cables, etc. 334 answered problems. 462pp. 5⅜ x 8½. 60754-2

MECHANICAL VIBRATIONS, J. P. Den Hartog. Classic textbook offers lucid explanations and illustrative models, applying theories of vibrations to a variety of practical industrial engineering problems. Numerous figures. 233 problems, solutions. Appendix. Index. Preface. 436pp. 5⅜ x 8½. 64785-4

STRENGTH OF MATERIALS, J. P. Den Hartog. Full, clear treatment of basic material (tension, torsion, bending, etc.) plus advanced material on engineering methods, applications. 350 answered problems. 323pp. 5⅜ x 8½. 60755-0

A HISTORY OF MECHANICS, René Dugas. Monumental study of mechanical principles from antiquity to quantum mechanics. Contributions of ancient Greeks, Galileo, Leonardo, Kepler, Lagrange, many others. 671pp. 5⅜ x 8½. 65632-2

METAL FATIGUE, N. E. Frost, K. J. Marsh, and L. P. Pook. Definitive, clearly written, and well-illustrated volume addresses all aspects of the subject, from the historical development of understanding metal fatigue to vital concepts of the cyclic stress that causes a crack to grow. Includes 7 appendixes. 544pp. 5⅜ x 8½. 40927-9

STATISTICAL MECHANICS: Principles and Applications, Terrell L. Hill. Standard text covers fundamentals of statistical mechanics, applications to fluctuation theory, imperfect gases, distribution functions, more. 448pp. 5⅜ x 8½. 65390-0

THE VARIATIONAL PRINCIPLES OF MECHANICS, Cornelius Lanczos. Graduate level coverage of calculus of variations, equations of motion, relativistic mechanics, more. First inexpensive paperbound edition of classic treatise. Index. Bibliography. 418pp. 5⅜ x 8½. 65067-7

THE VARIOUS AND INGENIOUS MACHINES OF AGOSTINO RAMELLI: A Classic Sixteenth-Century Illustrated Treatise on Technology, Agostino Ramelli. One of the most widely known and copied works on machinery in the 16th century. 194 detailed plates of water pumps, grain mills, cranes, more. 608pp. 9 x 12. 28180-9

ORDINARY DIFFERENTIAL EQUATIONS AND STABILITY THEORY: An Introduction, David A. Sánchez. Brief, modern treatment. Linear equation, stability theory for autonomous and nonautonomous systems, etc. 164pp. 5⅜ x 8¼. 63828-6

ROTARY WING AERODYNAMICS, W. Z. Stepniewski. Clear, concise text covers aerodynamic phenomena of the rotor and offers guidelines for helicopter performance evaluation. Originally prepared for NASA. 537 figures. 640pp. 6⅛ x 9¼. 64647-5

INTRODUCTION TO SPACE DYNAMICS, William Tyrrell Thomson. Comprehensive, classic introduction to space-flight engineering for advanced undergraduate and graduate students. Includes vector algebra, kinematics, transformation of coordinates. Bibliography. Index. 352pp. 5⅜ x 8½. 65113-4

HISTORY OF STRENGTH OF MATERIALS, Stephen P. Timoshenko. Excellent historical survey of the strength of materials with many references to the theories of elasticity and structure. 245 figures. 452pp. 5⅜ x 8½. 61187-6

ANALYTICAL FRACTURE MECHANICS, David J. Unger. Self-contained text supplements standard fracture mechanics texts by focusing on analytical methods for determining crack-tip stress and strain fields. 336pp. 6⅛ x 9¼. 41737-9

Mathematics

HANDBOOK OF MATHEMATICAL FUNCTIONS WITH FORMULAS, GRAPHS, AND MATHEMATICAL TABLES, edited by Milton Abramowitz and Irene A. Stegun. Vast compendium: 29 sets of tables, some to as high as 20 places. 1,046pp. 8 x 10½. 61272-4

FUNCTIONAL ANALYSIS (Second Corrected Edition), George Bachman and Lawrence Narici. Excellent treatment of subject geared toward students with background in linear algebra, advanced calculus, physics and engineering. Text covers introduction to inner-product spaces, normed, metric spaces, and topological spaces; complete orthonormal sets, the Hahn-Banach Theorem and its consequences, and many other related subjects. 1966 ed. 544pp. 6⅛ x 9¼. 40251-7

ASYMPTOTIC EXPANSIONS OF INTEGRALS, Norman Bleistein & Richard A. Handelsman. Best introduction to important field with applications in a variety of scientific disciplines. New preface. Problems. Diagrams. Tables. Bibliography. Index. 448pp. 5⅜ x 8½. 65082-0

FAMOUS PROBLEMS OF GEOMETRY AND HOW TO SOLVE THEM, Benjamin Bold. Squaring the circle, trisecting the angle, duplicating the cube: learn their history, why they are impossible to solve, then solve them yourself. 128pp. 5⅜ x 8½. 24297-8

VECTOR AND TENSOR ANALYSIS WITH APPLICATIONS, A. I. Borisenko and I. E. Tarapov. Concise introduction. Worked-out problems, solutions, exercises. 257pp. 5⅜ x 8¼. 63833-2

THE ABSOLUTE DIFFERENTIAL CALCULUS (CALCULUS OF TENSORS), Tullio Levi-Civita. Great 20th-century mathematician's classic work on material necessary for mathematical grasp of theory of relativity. 452pp. 5⅜ x 8¼. 63401-9

AN INTRODUCTION TO ORDINARY DIFFERENTIAL EQUATIONS, Earl A. Coddington. A thorough and systematic first course in elementary differential equations for undergraduates in mathematics and science, with many exercises and problems (with answers). Index. 304pp. 5⅜ x 8½. 65942-9

FOURIER SERIES AND ORTHOGONAL FUNCTIONS, Harry F. Davis. An incisive text combining theory and practical example to introduce Fourier series, orthogonal functions and applications of the Fourier method to boundary-value problems. 570 exercises. Answers and notes. 416pp. 5⅜ x 8½. 65973-9

COMPUTABILITY AND UNSOLVABILITY, Martin Davis. Classic graduate-level introduction to theory of computability, usually referred to as theory of recurrent functions. New preface and appendix. 288pp. 5⅜ x 8½. 61471-9

ASYMPTOTIC METHODS IN ANALYSIS, N. G. de Bruijn. An inexpensive, comprehensive guide to asymptotic methods–the pioneering work that teaches by explaining worked examples in detail. Index. 224pp. 5⅜ x 8½ 64221-6

ESSAYS ON THE THEORY OF NUMBERS, Richard Dedekind. Two classic essays by great German mathematician: on the theory of irrational numbers; and on transfinite numbers and properties of natural numbers. 115pp. 5⅜ x 8½. 21010-3

APPLIED COMPLEX VARIABLES, John W. Dettman. Step-by-step coverage of fundamentals of analytic function theory—plus lucid exposition of five important applications: Potential Theory; Ordinary Differential Equations; Fourier Transforms; Laplace Transforms; Asymptotic Expansions. 66 figures. Exercises at chapter ends. 512pp. 5⅜ x 8½. 64670-X

INTRODUCTION TO LINEAR ALGEBRA AND DIFFERENTIAL EQUA-TIONS, John W. Dettman. Excellent text covers complex numbers, determinants, orthonormal bases, Laplace transforms, much more. Exercises with solutions. Undergraduate level. 416pp. 5⅜ x 8½. 65191-6

MATHEMATICAL METHODS IN PHYSICS AND ENGINEERING, John W. Dettman. Algebraically based approach to vectors, mapping, diffraction, other topics in applied math. Also generalized functions, analytic function theory, more. Exercises. 448pp. 5⅜ x 8¼. 65649-7

CALCULUS OF VARIATIONS WITH APPLICATIONS, George M. Ewing. Applications-oriented introduction to variational theory develops insight and promotes understanding of specialized books, research papers. Suitable for advanced undergraduate/graduate students as primary, supplementary text. 352pp. 5⅜ x 8½. 64856-7

COMPLEX VARIABLES, Francis J. Flanigan. Unusual approach, delaying complex algebra till harmonic functions have been analyzed from real variable viewpoint. Includes problems with answers. 364pp. 5⅜ x 8½. 61388-7

AN INTRODUCTION TO THE CALCULUS OF VARIATIONS, Charles Fox. Graduate-level text covers variations of an integral, isoperimetrical problems, least action, special relativity, approximations, more. References. 279pp. 5⅜ x 8½. 65499-0

CATASTROPHE THEORY FOR SCIENTISTS AND ENGINEERS, Robert Gilmore. Advanced-level treatment describes mathematics of theory grounded in the work of Poincaré, R. Thom, other mathematicians. Also important applications to problems in mathematics, physics, chemistry and engineering. 1981 edition. References. 28 tables. 397 black-and-white illustrations. xvii + 666pp. 6⅛ x 9¼. 67539-4

INTRODUCTION TO DIFFERENCE EQUATIONS, Samuel Goldberg. Exceptionally clear exposition of important discipline with applications to sociology, psychology, economics. Many illustrative examples; over 250 problems. 260pp. 5⅜ x 8½. 65084-7

NUMERICAL METHODS FOR SCIENTISTS AND ENGINEERS, Richard Hamming. Classic text stresses frequency approach in coverage of algorithms, polynomial approximation, Fourier approximation, exponential approximation, other topics. Revised and enlarged 2nd edition. 721pp. 5⅜ x 8½. 65241-6

INTRODUCTION TO NUMERICAL ANALYSIS (2nd Edition), F. B. Hildebrand. Classic, fundamental treatment covers computation, approximation, interpolation, numerical differentiation and integration, other topics. 150 new problems. 669pp. 5⅜ x 8½. 65363-3

THE FUNCTIONS OF MATHEMATICAL PHYSICS, Harry Hochstadt. Comprehensive treatment of orthogonal polynomials, hypergeometric functions, Hill's equation, much more. Bibliography. Index. 322pp. 5⅜ x 8½. 65214-9

THREE PEARLS OF NUMBER THEORY, A. Y. Khinchin. Three compelling puzzles require proof of a basic law governing the world of numbers. Challenges concern van der Waerden's theorem, the Landau-Schnirelmann hypothesis and Mann's theorem, and a solution to Waring's problem. Solutions included. 64pp. 5⅜ x 8¼. 40026-3

CALCULUS REFRESHER FOR TECHNICAL PEOPLE, A. Albert Klaf. Covers important aspects of integral and differential calculus via 756 questions. 566 problems, most answered. 431pp. 5⅜ x 8½. 20370-0

THE PHILOSOPHY OF MATHEMATICS: An Introductory Essay, Stephan Körner. Surveys the views of Plato, Aristotle, Leibniz & Kant concerning propositions and theories of applied and pure mathematics. Introduction. Two appendices. Index. 198pp. 5⅜ x 8½. 25048-2

INTRODUCTORY REAL ANALYSIS, A.N. Kolmogorov, S. V. Fomin. Translated by Richard A. Silverman. Self-contained, evenly paced introduction to real and functional analysis. Some 350 problems. 403pp. 5⅜ x 8½. 61226-0

APPLIED ANALYSIS, Cornelius Lanczos. Classic work on analysis and design of finite processes for approximating solution of analytical problems. Algebraic equations, matrices, harmonic analysis, quadrature methods, much more. 559pp. 5⅜ x 8½. 65656-X

AN INTRODUCTION TO ALGEBRAIC STRUCTURES, Joseph Landin. Superb self-contained text covers "abstract algebra": sets and numbers, theory of groups, theory of rings, much more. Numerous well-chosen examples, exercises. 247pp. 5⅜ x 8½. 65940-2

SPECIAL FUNCTIONS, N. N. Lebedev. Translated by Richard Silverman. Famous Russian work treating more important special functions, with applications to specific problems of physics and engineering. 38 figures. 308pp. 5⅜ x 8½. 60624-4

QUALITATIVE THEORY OF DIFFERENTIAL EQUATIONS, V. V. Nemytskii and V.V. Stepanov. Classic graduate-level text by two prominent Soviet mathematicians covers classical differential equations as well as topological dynamics and ergodic theory. Bibliographies. 523pp. 5⅜ x 8½. 65954-2

NUMBER THEORY AND ITS HISTORY, Oystein Ore. Unusually clear, accessible introduction covers counting, properties of numbers, prime numbers, much more. Bibliography. 380pp. 5⅜ x 8½. 65620-9

THEORY OF MATRICES, Sam Perlis. Outstanding text covering rank, nonsingularity and inverses in connection with the development of canonical matrices under the relation of equivalence, and without the intervention of determinants. Includes exercises. 237pp. 5⅜ x 8½. 66810-X

INTRODUCTION TO ANALYSIS, Maxwell Rosenlicht. Unusually clear, accessible coverage of set theory, real number system, metric spaces, continuous functions, Riemann integration, multiple integrals, more. Wide range of problems. Undergraduate level. Bibliography. 254pp. 5⅜ x 8½. 65038-3

MODERN NONLINEAR EQUATIONS, Thomas L. Saaty. Emphasizes practical solution of problems; covers seven types of equations. ". . . a welcome contribution to the existing literature...."–*Math Reviews.* 490pp. 5⅜ x 8½. 64232-1

MATRICES AND LINEAR ALGEBRA, Hans Schneider and George Phillip Barker. Basic textbook covers theory of matrices and its applications to systems of linear equations and related topics such as determinants, eigenvalues and differential equations. Numerous exercises. 432pp. 5⅜ x 8½. 66014-1

MATHEMATICS APPLIED TO CONTINUUM MECHANICS, Lee A. Segel. Analyzes models of fluid flow and solid deformation. For upper-level math, science and engineering students. 608pp. 5⅜ x 8½. 65369-2

ELEMENTS OF REAL ANALYSIS, David A. Sprecher. Classic text covers fundamental concepts, real number system, point sets, functions of a real variable, Fourier series, much more. Over 500 exercises. 352pp. 5⅜ x 8½. 65385-4

AN INTRODUCTION TO MATRICES, SETS AND GROUPS FOR SCIENCE STUDENTS, G. Stephenson. Concise, readable text introduces sets, groups, and most importantly, matrices to undergraduate students of physics, chemistry, and engineering. Problems. 164pp. 5⅜ x 8½. 65077-4

SET THEORY AND LOGIC, Robert R. Stoll. Lucid introduction to unified theory of mathematical concepts. Set theory and logic seen as tools for conceptual understanding of real number system. 496pp. 5⅜ x 8¼. 63829-4

TENSOR CALCULUS, J.L. Synge and A. Schild. Widely used introductory text covers spaces and tensors, basic operations in Riemannian space, non-Riemannian spaces, etc. 324pp. 5⅜ x 8¼. 63612-7

ORDINARY DIFFERENTIAL EQUATIONS, Morris Tenenbaum and Harry Pollard. Exhaustive survey of ordinary differential equations for undergraduates in mathematics, engineering, science. Thorough analysis of theorems. Diagrams. Bibliography. Index. 818pp. 5⅜ x 8½. 64940-7

INTEGRAL EQUATIONS, F. G. Tricomi. Authoritative, well-written treatment of extremely useful mathematical tool with wide applications. Volterra Equations, Fredholm Equations, much more. Advanced undergraduate to graduate level. Exercises. Bibliography. 238pp. 5⅜ x 8½. 64828-1

FOURIER SERIES, Georgi P. Tolstov. Translated by Richard A. Silverman. A valuable addition to the literature on the subject, moving clearly from subject to subject and theorem to theorem. 107 problems, answers. 336pp. 5⅜ x 8½. 63317-9

POPULAR LECTURES ON MATHEMATICAL LOGIC, Hao Wang. Noted logician's lucid treatment of historical developments, set theory, model theory, recursion theory and constructivism, proof theory, more. 3 appendixes. Bibliography. 1981 edition. ix + 283pp. 5⅜ x 8½. 67632-3

CALCULUS OF VARIATIONS, Robert Weinstock. Basic introduction covering isoperimetric problems, theory of elasticity, quantum mechanics, electrostatics, etc. Exercises throughout. 326pp. 5⅜ x 8½. 63069-2

THE CONTINUUM: A Critical Examination of the Foundation of Analysis, Hermann Weyl. Classic of 20th-century foundational research deals with the conceptual problem posed by the continuum. 156pp. 5⅜ x 8½. 67982-9

CHALLENGING MATHEMATICAL PROBLEMS WITH ELEMENTARY SOLUTIONS, A. M. Yaglom and I. M. Yaglom. Over 170 challenging problems on probability theory, combinatorial analysis, points and lines, topology, convex polygons, many other topics. Solutions. Total of 445pp. 5⅜ x 8½. Two-vol. set. Vol. I: 65536-9 Vol. II: 65537-7

A SURVEY OF NUMERICAL MATHEMATICS, David M. Young and Robert Todd Gregory. Broad self-contained coverage of computer-oriented numerical algorithms for solving various types of mathematical problems in linear algebra, ordinary and partial, differential equations, much more. Exercises. Total of 1,248pp. 5⅜ x 8½. Two volumes. Vol. I: 65691-8 Vol. II: 65692-6

INTRODUCTION TO PARTIAL DIFFERENTIAL EQUATIONS WITH APPLICATIONS, E. C. Zachmanoglou and Dale W. Thoe. Essentials of partial differential equations applied to common problems in engineering and the physical sciences. Problems and answers. 416pp. 5⅜ x 8½. 65251-3

THE THEORY OF GROUPS, Hans J. Zassenhaus. Well-written graduate-level text acquaints reader with group-theoretic methods and demonstrates their usefulness in mathematics. Axioms, the calculus of complexes, homomorphic mapping, p-group theory, more. Many proofs shorter and more transparent than older ones. 276pp. 5⅜ x 8½. 40922-8

DISTRIBUTION THEORY AND TRANSFORM ANALYSIS: An Introduction to Generalized Functions, with Applications, A. H. Zemanian. Provides basics of distribution theory, describes generalized Fourier and Laplace transformations. Numerous problems. 384pp. 5⅜ x 8½. 65479-6

Math–Decision Theory, Statistics, Probability

ELEMENTARY DECISION THEORY, Herman Chernoff and Lincoln E. Moses. Clear introduction to statistics and statistical theory covers data processing, probability and random variables, testing hypotheses, much more. Exercises. 364pp. 5⅜ x 8½. 65218-1

STATISTICS MANUAL, Edwin L. Crow et al. Comprehensive, practical collection of classical and modern methods prepared by U.S. Naval Ordnance Test Station. Stress on use. Basics of statistics assumed. 288pp. 5⅜ x 8½. 60599-X

SOME THEORY OF SAMPLING, William Edwards Deming. Analysis of the problems, theory and design of sampling techniques for social scientists, industrial managers and others who find statistics important at work. 61 tables. 90 figures. xvii +602pp. 5⅜ x 8½. 64684-X

STATISTICAL ADJUSTMENT OF DATA, W. Edwards Deming. Introduction to basic concepts of statistics, curve fitting, least squares solution, conditions without parameter, conditions containing parameters. 26 exercises worked out. 271pp. 5⅜ x 8½. 64685-8

LINEAR PROGRAMMING AND ECONOMIC ANALYSIS, Robert Dorfman, Paul A. Samuelson and Robert M. Solow. First comprehensive treatment of linear programming in standard economic analysis. Game theory, modern welfare economics, Leontief input-output, more. 525pp. 5⅜ x 8½. 65491-5

DICTIONARY/OUTLINE OF BASIC STATISTICS, John E. Freund and Frank J. Williams. A clear concise dictionary of over 1,000 statistical terms and an outline of statistical formulas covering probability, nonparametric tests, much more. 208pp. 5⅜ x 8½. 66796-0

PROBABILITY: An Introduction, Samuel Goldberg. Excellent basic text covers set theory, probability theory for finite sample spaces, binomial theorem, much more. 360 problems. Bibliographies. 322pp. 5⅜ x 8½. 65252-1

GAMES AND DECISIONS: Introduction and Critical Survey, R. Duncan Luce and Howard Raiffa. Superb nontechnical introduction to game theory, primarily applied to social sciences. Utility theory, zero-sum games, n-person games, decision-making, much more. Bibliography. 509pp. 5⅜ x 8½. 65943-7

FIFTY CHALLENGING PROBLEMS IN PROBABILITY WITH SOLUTIONS, Frederick Mosteller. Remarkable puzzlers, graded in difficulty, illustrate elementary and advanced aspects of probability. Detailed solutions. 88pp. 5⅜ x 8½. 65355-2

PROBABILITY THEORY: A Concise Course, Y. A. Rozanov. Highly readable, self-contained introduction covers combination of events, dependent events, Bernoulli trials, etc. 148pp. 5⅜ x 8¼. 63544-9

STATISTICAL METHOD FROM THE VIEWPOINT OF QUALITY CONTROL, Walter A. Shewhart. Important text explains regulation of variables, uses of statistical control to achieve quality control in industry, agriculture, other areas. 192pp. 5⅜ x 8½. 65232-7

THE COMPLEAT STRATEGYST: Being a Primer on the Theory of Games of Strategy, J. D. Williams. Highly entertaining classic describes, with many illustrated examples, how to select best strategies in conflict situations. Prefaces. Appendices. 268pp. 5⅜ x 8½. 25101-2

Math–Geometry and Topology

ELEMENTARY CONCEPTS OF TOPOLOGY, Paul Alexandroff. Elegant, intuitive approach to topology from set-theoretic topology to Betti groups; how concepts of topology are useful in math and physics. 25 figures. 57pp. 5⅜ x 8½. 60747-X

COMBINATORIAL TOPOLOGY, P. S. Alexandrov. Clearly written, well-organized, three-part text begins by dealing with certain classic problems without using the formal techniques of homology theory and advances to the central concept, the Betti groups. Numerous detailed examples. 654pp. 5⅜ x 8½. 40179-0

EXPERIMENTS IN TOPOLOGY, Stephen Barr. Classic, lively explanation of one of the byways of mathematics. Klein bottles, Moebius strips, projective planes, map coloring, problem of the Koenigsberg bridges, much more, described with clarity and wit. 43 figures. 210pp. 5⅜ x 8½. 25933-1

CONFORMAL MAPPING ON RIEMANN SURFACES, Harvey Cohn. Lucid, insightful book presents ideal coverage of subject. 334 exercises make book perfect for self-study. 55 figures. 352pp. 5⅝ x 8¼. 64025-6

THE GEOMETRY OF RENÉ DESCARTES, René Descartes. The great work founded analytical geometry. Original French text, Descartes's own diagrams, together with definitive Smith-Latham translation. 244pp. 5⅜ x 8½. 60068-8

THE THIRTEEN BOOKS OF EUCLID'S ELEMENTS, translated with introduction and commentary by Sir Thomas L. Heath. Definitive edition. Textual and linguistic notes, mathematical analysis. 2,500 years of critical commentary. Unabridged. 1,414pp. 5⅜ x 8½. Three-vol. set.
Vol. I: 60088-2 Vol. II: 60089-0 Vol. III: 60090-4

GEOMETRY OF COMPLEX NUMBERS, Hans Schwerdtfeger. Illuminating, widely praised book on analytic geometry of circles, the Moebius transformation, and two-dimensional non-Euclidean geometries. 200pp. 5⅝ x 8¼. 63830-8

DIFFERENTIAL GEOMETRY, Heinrich W. Guggenheimer. Local differential geometry as an application of advanced calculus and linear algebra. Curvature, transformation groups, surfaces, more. Exercises. 62 figures. 378pp. 5⅜ x 8½. 63433-7

CURVATURE AND HOMOLOGY: Enlarged Edition, Samuel I. Goldberg. Revised edition examines topology of differentiable manifolds; curvature, homology of Riemannian manifolds; compact Lie groups; complex manifolds; curvature, homology of Kaehler manifolds. New Preface. Four new appendixes. 416pp. 5⅜ x 8½. 40207-X

TOPOLOGY, John G. Hocking and Gail S. Young. Superb one-year course in classical topology. Topological spaces and functions, point-set topology, much more. Examples and problems. Bibliography. Index. 384pp. 5⅝ x 8¼. 65676-4

LECTURES ON CLASSICAL DIFFERENTIAL GEOMETRY, Second Edition, Dirk J. Struik. Excellent brief introduction covers curves, theory of surfaces, fundamental equations, geometry on a surface, conformal mapping, other topics. Problems. 240pp. 5⅜ x 8½. 65609-8

Math–History of

A SHORT ACCOUNT OF THE HISTORY OF MATHEMATICS, W. W. Rouse Ball. One of clearest, most authoritative surveys from the Egyptians and Phoenicians through 19th-century figures such as Grassman, Galois, Riemann. Fourth edition. 522pp. 5⅜ x 8½. 20630-0

THE HISTORY OF THE CALCULUS AND ITS CONCEPTUAL DEVELOPMENT, Carl B. Boyer. Origins in antiquity, medieval contributions, work of Newton, Leibniz, rigorous formulation. Treatment is verbal. 346pp. 5⅜ x 8½. 60509-4

THE HISTORICAL ROOTS OF ELEMENTARY MATHEMATICS, Lucas N. H. Bunt, Phillip S. Jones, and Jack D. Bedient. Fundamental underpinnings of modern arithmetic, algebra, geometry and number systems derived from ancient civilizations. 320pp. 5⅜ x 8½. 25563-8

A HISTORY OF MATHEMATICAL NOTATIONS, Florian Cajori. This classic study notes the first appearance of a mathematical symbol and its origin, the competition it encountered, its spread among writers in different countries, its rise to popularity, its eventual decline or ultimate survival. Original 1929 two-volume edition presented here in one volume. xxviii+820pp. 5⅜ x 8½. 67766-4

GAMES, GODS & GAMBLING: A History of Probability and Statistical Ideas, F. N. David. Episodes from the lives of Galileo, Fermat, Pascal, and others illustrate this fascinating account of the roots of mathematics. Features thought-provoking references to classics, archaeology, biography, poetry. 1962 edition. 304pp. 5⅜ x 8½. (Available in U.S. only.) 40023-9

OF MEN AND NUMBERS: The Story of the Great Mathematicians, Jane Muir. Fascinating accounts of the lives and accomplishments of history's greatest mathematical minds–Pythagoras, Descartes, Euler, Pascal, Cantor, many more. Anecdotal, illuminating. 30 diagrams. Bibliography. 256pp. 5⅜ x 8½. 28973-7

HISTORY OF MATHEMATICS, David E. Smith. Nontechnical survey from ancient Greece and Orient to late 19th century; evolution of arithmetic, geometry, trigonometry, calculating devices, algebra, the calculus. 362 illustrations. 1,355pp. 5⅜ x 8½. Two-vol. set. Vol. I: 20429-4 Vol. II: 20430-8

A CONCISE HISTORY OF MATHEMATICS, Dirk J. Struik. The best brief history of mathematics. Stresses origins and covers every major figure from ancient Near East to 19th century. 41 illustrations. 195pp. 5⅜ x 8½. 60255-9

Physics

OPTICAL RESONANCE AND TWO-LEVEL ATOMS, L. Allen and J. H. Eberly. Clear, comprehensive introduction to basic principles behind all quantum optical resonance phenomena. 53 illustrations. Preface. Index. 256pp. 5⅜ x 8½.   65533-4

ULTRASONIC ABSORPTION: An Introduction to the Theory of Sound Absorption and Dispersion in Gases, Liquids and Solids, A. B. Bhatia. Standard reference in the field provides a clear, systematically organized introductory review of fundamental concepts for advanced graduate students, research workers. Numerous diagrams. Bibliography. 440pp. 5⅜ x 8½.   64917-2

QUANTUM THEORY, David Bohm. This advanced undergraduate-level text presents the quantum theory in terms of qualitative and imaginative concepts, followed by specific applications worked out in mathematical detail. Preface. Index. 655pp. 5⅜ x 8½.   65969-0

ATOMIC PHYSICS (8th edition), Max Born. Nobel laureate's lucid treatment of kinetic theory of gases, elementary particles, nuclear atom, wave-corpuscles, atomic structure and spectral lines, much more. Over 40 appendices, bibliography. 495pp. 5⅜ x 8½.   65984-4

AN INTRODUCTION TO HAMILTONIAN OPTICS, H. A. Buchdahl. Detailed account of the Hamiltonian treatment of aberration theory in geometrical optics. Many classes of optical systems defined in terms of the symmetries they possess. Problems with detailed solutions. 1970 edition. xv + 360pp. 5⅜ x 8½.   67597-1

THIRTY YEARS THAT SHOOK PHYSICS: The Story of Quantum Theory, George Gamow. Lucid, accessible introduction to influential theory of energy and matter. Careful explanations of Dirac's anti-particles, Bohr's model of the atom, much more. 12 plates. Numerous drawings. 240pp. 5⅜ x 8½.   24895-X

ELECTRONIC STRUCTURE AND THE PROPERTIES OF SOLIDS: The Physics of the Chemical Bond, Walter A. Harrison. Innovative text offers basic understanding of the electronic structure of covalent and ionic solids, simple metals, transition metals and their compounds. Problems. 1980 edition. 582pp. 6⅛ x 9¼.   66021-4

HYDRODYNAMIC AND HYDROMAGNETIC STABILITY, S. Chandrasekhar. Lucid examination of the Rayleigh-Benard problem; clear coverage of the theory of instabilities causing convection. 704pp. 5⅜ x 8¼.   64071-X

INVESTIGATIONS ON THE THEORY OF THE BROWNIAN MOVEMENT, Albert Einstein. Five papers (1905–8) investigating dynamics of Brownian motion and evolving elementary theory. Notes by R. Fürth. 122pp. 5⅜ x 8½.   60304-0

THE PHYSICS OF WAVES, William C. Elmore and Mark A. Heald. Unique overview of classical wave theory. Acoustics, optics, electromagnetic radiation, more. Ideal as classroom text or for self-study. Problems. 477pp. 5⅜ x 8½.   64926-1

PHYSICAL PRINCIPLES OF THE QUANTUM THEORY, Werner Heisenberg. Nobel Laureate discusses quantum theory, uncertainty, wave mechanics, work of Dirac, Schroedinger, Compton, Wilson, Einstein, etc. 184pp. 5⅜ x 8½.   60113-7

ATOMIC SPECTRA AND ATOMIC STRUCTURE, Gerhard Herzberg. One of best introductions; especially for specialist in other fields. Treatment is physical rather than mathematical. 80 illustrations. 257pp. 5⅜ x 8½.   60115-3

AN INTRODUCTION TO STATISTICAL THERMODYNAMICS, Terrell L. Hill. Excellent basic text offers wide-ranging coverage of quantum statistical mechanics, systems of interacting molecules, quantum statistics, more. 523pp. 5⅜ x 8½.
65242-4

THEORETICAL PHYSICS, Georg Joos, with Ira M. Freeman. Classic overview covers essential math, mechanics, electromagnetic theory, thermodynamics, quantum mechanics, nuclear physics, other topics. First paperback edition. xxiii + 885pp. 5⅜ x 8½.   65227-0

PROBLEMS AND SOLUTIONS IN QUANTUM CHEMISTRY AND PHYSICS, Charles S. Johnson, Jr. and Lee G. Pedersen. Unusually varied problems, detailed solutions in coverage of quantum mechanics, wave mechanics, angular momentum, molecular spectroscopy, more. 280 problems plus 139 supplementary exercises. 430pp. 6½ x 9¼.   65236-X

THEORETICAL SOLID STATE PHYSICS, Vol. 1: Perfect Lattices in Equilibrium; Vol. II: Non-Equilibrium and Disorder, William Jones and Norman H. March. Monumental reference work covers fundamental theory of equilibrium properties of perfect crystalline solids, non-equilibrium properties, defects and disordered systems. Appendices. Problems. Preface. Diagrams. Index. Bibliography. Total of 1,301pp. 5⅜ x 8½. Two volumes.   Vol. I: 65015-4   Vol. II: 65016-2

A TREATISE ON ELECTRICITY AND MAGNETISM, James Clerk Maxwell. Important foundation work of modern physics. Brings to final form Maxwell's theory of electromagnetism and rigorously derives his general equations of field theory. 1,084pp. 5⅜ x 8½. Two-vol. set.   Vol. I: 60636-8   Vol. II: 60637-6

OPTICKS, Sir Isaac Newton. Newton's own experiments with spectroscopy, colors, lenses, reflection, refraction, etc., in language the layman can follow. Foreword by Albert Einstein. 532pp. 5⅜ x 8½.   60205-2

THEORY OF ELECTROMAGNETIC WAVE PROPAGATION, Charles Herach Papas. Graduate-level study discusses the Maxwell field equations, radiation from wire antennas, the Doppler effect and more. xiii + 244pp. 5⅜ x 8½.   65678-5

INTRODUCTION TO QUANTUM MECHANICS With Applications to Chemistry, Linus Pauling & E. Bright Wilson, Jr. Classic undergraduate text by Nobel Prize winner applies quantum mechanics to chemical and physical problems. Numerous tables and figures enhance the text. Chapter bibliographies. Appendices. Index. 468pp. 5⅜ x 8½.   64871-0

METHODS OF THERMODYNAMICS, Howard Reiss. Outstanding text focuses on physical technique of thermodynamics, typical problem areas of understanding, and significance and use of thermodynamic potential. 1965 edition. 238pp. 5⅜ x 8½. 69445-3

TENSOR ANALYSIS FOR PHYSICISTS, J. A. Schouten. Concise exposition of the mathematical basis of tensor analysis, integrated with well-chosen physical examples of the theory. Exercises. Index. Bibliography. 289pp. 5⅜ x 8½. 65582-2

RELATIVITY IN ILLUSTRATIONS, Jacob T. Schwartz. Clear nontechnical treatment makes relativity more accessible than ever before. Over 60 drawings illustrate concepts more clearly than text alone. Only high school geometry needed. Bibliography. 128pp. 6⅛ x 9¼. 25965-X

THE ELECTROMAGNETIC FIELD, Albert Shadowitz. Comprehensive undergraduate text covers basics of electric and magnetic fields, builds up to electromagnetic theory. Also related topics, including relativity. Over 900 problems. 768pp. 5⅝ x 8¼. 65660-8

GREAT EXPERIMENTS IN PHYSICS: Firsthand Accounts from Galileo to Einstein, edited by Morris H. Shamos. 25 crucial discoveries: Newton's laws of motion, Chadwick's study of the neutron, Hertz on electromagnetic waves, more. Original accounts clearly annotated. 370pp. 5⅜ x 8½. 25346-5

RELATIVITY, THERMODYNAMICS AND COSMOLOGY, Richard C. Tolman. Landmark study extends thermodynamics to special, general relativity; also applications of relativistic mechanics, thermodynamics to cosmological models. 501pp. 5⅜ x 8½. 65383-8

LIGHT SCATTERING BY SMALL PARTICLES, H. C. van de Hulst. Comprehensive treatment including full range of useful approximation methods for researchers in chemistry, meteorology and astronomy. 44 illustrations. 470pp. 5⅜ x 8½. 64228-3

STATISTICAL PHYSICS, Gregory H. Wannier. Classic text combines thermodynamics, statistical mechanics and kinetic theory in one unified presentation of thermal physics. Problems with solutions. Bibliography. 532pp. 5⅜ x 8½. 65401-X

Paperbound unless otherwise indicated. Available at your book dealer, online at **www.doverpublications.com**, or by writing to Dept. GI, Dover Publications, Inc., 31 East 2nd Street, Mineola, NY 11501. For current price information or for free catalogues (please indicate field of interest), write to Dover Publications or log on to **www.doverpublications.com** and see every Dover book in print. Dover publishes more than 500 books each year on science, elementary and advanced mathematics, biology, music, art, literary history, social sciences, and other areas.